The Electronic Frontier

Navigating the Evolving World of Technology and Innovation in the Digital Age for a Brighter Future

NAGINDERPAL SINGH

DEDICATED TO

I am dedicating this book to my parents Sardar Narinder Singh and Late Sardarni Gurjeet Kaur

FOREWORD

In the world of human development, few factors have been more disruptive as the fast growth of technology. From the genesis of the World Wide Web to the advent of artificial intelligence, we are witness to an age that redefines the frontiers of human power and connectedness. "The Electronic Frontier: Navigating the Evolving World of Technology and Innovation in the Digital Age for a Brighter Future" is not simply a book; it is a compass for navigating this new region.

In these pages, you will go on a voyage across the digital world, led by a thorough analysis of the numerous features that characterize our technology era. From the historical foundations of the digital revolution to the ethical questions defining our future, this book presents a panoramic picture of the electronic frontier. It exceeds ordinary explanation, presenting insights and comments that urge us to ponder not just what technology can achieve, but what it should do.

What sets this work distinctive is its persistent dedication to the human dimension. While it goes deep into the subtleties of algorithms, networks, and artificial intelligence, it never loses sight of the persons who produce, interact, and are influenced by these advances. Through colorful tales and rigorous analysis, it shows the deep ways in which technology impacts our society, economy, and personal lives.

The essence of this book rests in its goal for a greater digital future. It compels us to picture a future where technology advancement harmonizes with human values - a society where inclusion, sustainability, and ethics are not just goals, but intrinsic components of innovation. Each chapter unfolds like a blueprint, revealing insights into how we may collaboratively construct this future, one choice, one invention, at a time.

As you begin on this intellectual voyage, I invite you to approach each page not only as a reader, but as an active participant in the continuous story of the digital era. Engage with the concepts, question the assumptions, and, above all, let this book to inspire you to create the technological frontier in a manner that empowers and uplifts us all.

"The Electronic Frontier" is not a forecast; it is an invitation. An invitation to contemplate what is feasible, what is desired, and what is required in this quickly shifting day. It reminds us that although technology is a wonder of human ingenuity, it is the human spirit that must ultimately drive its course.

Let this book serve as a guide on your voyage into the technological frontier. May it excite you, challenge you, and eventually encourage you to contribute to a digital future that represents the very best of our common humanity.

Narinder Singh

PREFACE

In an age marked by the fast advancement of technology, we find ourselves standing at the edge of a digital frontier, ready to design a future that both unlimited potential and deep difficulties. This book, "The Electronic Frontier: Navigating the Evolving World of Technology and Innovation in the Digital Age for a Brighter Future," is an investigation of this dynamic terrain.

Within these pages, we begin on a trip across the diverse worlds of the digital era. We dig into the historical currents that have carried us into this new age, following the strands of invention, communication, and social transformation. Through each chapter, we strive to illustrate not just the remarkable achievements that have altered our world but also the ethical imperatives that must follow them.

This book is not a simple catalog of technical miracles, but a summons to contemplation and action. It encourages us to evaluate the repercussions of our technological choices, to defend the values that define our humanity, and to create inclusion in a digital universe that knows no borders. It is an invitation to think critically, to engage with the intricacies of our digital life, and to envisage a future where technology is a force for social good.

As we navigate through chapters that explore the internet's vast ecosystem, the transformative power of information technology, and the profound societal shifts driven by artificial intelligence, we are guided by a central conviction: that the potential for a brighter digital future lies within our grasp. By adopting ethical ideals, supporting inclusion, and emphasizing sustainability, we can design a future where technology serves as a light of progress, enhancing the lives of people and communities.

In publishing this book, our purpose is to give a compass for individuals who desire to traverse this technological frontier. It is an offering to students, academics, innovators, policymakers, and everyone who interact with the digital domain, asking them to go on a journey of investigation, reflection, and empowerment.

As we start on this collaborative project, we welcome you, dear reader, to join us in imagining and constructing a future where the electronic frontier is a source of education, empowerment, and a better tomorrow. Together, let us pave a path towards a digital era that represents the very best of our shared human potential.

With optimism for a future governed by knowledge, empathy, and creativity,

Naginderpal Singh

ACKNOWLEDGMENTS

This book would not have been possible without the everlasting support and affection of my parents, Narinder Singh and Late Gurjeet Kaur. Their impact has been the cornerstone upon which our enterprise rests.

To my father, Narinder Singh, your conviction in my skills has been a guiding light throughout my journey. Your knowledge, support, and continual presence have formed not just this book but also the person I have become. Your unfailing support has been my source of strength.

To my late mother, Gurjeet Kaur, you remain an eternal inspiration. Your memory continues to be a light of love and drive, inspiring me to strive for perfection. Your legacy lives on in every word penned inside these pages.

Together, you have been my pillars of strength, my source of unconditional love, and my greatest role models. Your efforts, devotion, and everlasting confidence in my ambitions have been the driving factor behind my success.

I also express my deepest appreciation to my extended family and friends who have been by my side, giving support and understanding during this journey. Your trust in me has been tremendous.

This book is a testimonial to the deep influence of your love and support. It serves as a testimony to the principles you instilled in me and the unshakeable confidence you had in my ability. Thank you, from the bottom of my heart, for being my biggest advocates.

With love and appreciation,

Naginderpal Singh

CONTENTS

Dedication   *iii*
Foreword   *v*
Preface   *viii*
Acknowledgement   *xi*

1. The Digital Revolution and its Impacts   15

2. The Internet and its Ecosystem   30

3. Innovations in Information Technology   45

4. The Digital Economy and Entrepreneurship   61

5. Privacy and Data Protection   76

6. Artificial Intelligence and Society   90

7. Cyber Security and Digital Threats   105

8. Digital Inclusion and Accessibility   121

9. Environmental Sustainability and Technology   137

10. The Future of Technology and Human Society   153

Chapter 1: The Digital Revolution and Its Impacts

1.1. The Historical Context of the Digital Age

"The Historical Context of the Digital Age" involves a fundamental transition in human civilization, typified by the development and expansion of digital technology. This epochal change traces its origins back to the mid-20th century, with the emergence of the digital computer. Initially, these computers were gigantic, sophisticated, and dedicated for specific duties within academic, military, and political realms. The Digital Age fully took flight in the second part of the 20th century, pushed by discoveries in microelectronics and integrated circuits, which led to the shrinking and popularization of computer equipment. This shift was highlighted by significant milestones like the creation of the microprocessor by Intel in 1971, ushering a new age of personal computing. The Digital Age's basis was therefore constructed upon a sequence of technical advancements, increasingly democratizing access to computer power.

As the Digital Age gathered speed, it started penetrating numerous sectors of human existence. The development of the Internet in the late 20th century might be regarded a watershed point. The Internet, first designed as a communication tool for military and scholarly reasons, soon grew into a worldwide network linking people, information, and systems around the globe. This move democratized access to information, altering how knowledge is transmitted, shared, and consumed. The advent of the World Wide Web in 1991 by Tim Berners-Lee further hastened this trend, allowing users to browse the immense expanse of digital material with ease. Consequently, the Internet became a fundamental part of contemporary life, transforming how people communicate, connect, and do business.

Furthermore, the Digital Age has dramatically altered economies and businesses on a worldwide basis. The integration of digital technology in numerous sectors led to the automation of operations, enhanced efficiency, and the emergence of totally new industries. E-commerce platforms evolved, disrupting old retail patterns and opening the door for a global economy. Additionally, sectors such as entertainment and media saw profound upheavals, with digital platforms changing content generation, delivery, and consumption. The music and film industries, for instance, saw a paradigm shift when digital formats superseded physical media, giving birth to new means of content distribution.

Society's cultural fabric also witnessed major transformation in the aftermath of the Digital Age. The expansion of social media platforms and online groups generated new kinds of social interaction and identity expression. Individuals and organizations now have unparalleled tools to communicate, exchange experiences, and organize around numerous causes. The Digital Age has also created outlets for disadvantaged voices and movements, democratizing the potential to influence public debate.

In conclusion, the historical setting of the Digital Age is founded in a sequence of technical advancements, extending from the creation of early computers to the worldwide ubiquity of digital technology in the 21st century. This change has fundamentally transformed how people communicate, acquire information, do business, and engage with one another. It symbolizes a key moment in human history, analogous to past industrial revolutions, with far-reaching repercussions for society, economics, and culture. As we continue to cross this digital frontier, recognizing its historical origins is vital for appreciating the multiple affects technology imposes on our lives.

1.2. The Transformation of Communication

"The Transformation of Communication" refers to the dramatic development that the area of communication has experienced with the arrival of digital technology. This change has affected the way people, groups, and civilizations connect, share information, and express themselves. One of the most crucial parts of this revolution is the move from conventional, one-way communication channels to a dynamic, multi-directional interchange of information.

In the pre-digital period, communication mostly happened via traditional channels such as newspapers, radio, and television. These platforms were typified by a top-down approach, where a small handful of gatekeepers controlled the flow of information to the masses. Information moved in a unidirectional way, from the source to the viewer, without much possibility for interaction or comment.

This paradigm restricted individual engagement and confined the function of the audience to that of passive receivers.

The digital era, however, has ushered in a democratization of communication. The expansion of the internet, social media, and other digital platforms has allowed people to become both makers and consumers of information. Social media platforms, in particular, have developed as strong instruments for self-expression, allowing users to share their ideas, views, and experiences with a worldwide audience. This transformation has profoundly transformed the nature of communication, allowing for a more inclusive, varied, and interactive dialogue.

Furthermore, the revolution of communication has also led to an unparalleled amount of connectedness. Geographical and temporal borders have become more meaningless since individuals may now connect in real-time throughout the world. This interconnectedness has aided the quick spread of information, allowing news and events to reach a worldwide audience virtually immediately. It has also promoted the establishment of virtual communities and networks, where people with common interests may interact and cooperate regardless of physical proximity.

However, it is crucial to note that this shift is not without its hurdles. The digital age has brought forth worries about the authenticity and dependability of information. The ease with which material may be made and spread has led to concerns such as disinformation, false news, and online echo chambers. Discerning the authenticity of sources has become a key skill in this information-rich era.

In conclusion, the change of communication in the digital age reflects a paradigm shift in how information is created, transmitted, and consumed. It has democratized the process, allowing for more inclusion and involvement in the dialogue. Additionally, it has produced unprecedented levels of global connection. However, this transition also raises issues in terms of information validity and reliability. Navigating this dynamic terrain demands a critical awareness of the digital communication ecology and the capacity to pick reputable sources from a flood of information.

1.3. Digital Disruption in Traditional Industries

"Digital Disruption in conventional Industries" refers to the revolutionary influence of technology on established sectors of the economy that have traditionally functioned in conventional ways. This phenomena is defined by the emergence of digital technology and creative business models that challenge and frequently disrupt traditional practices. One of the major instances of digital disruption may be observed in the retail business. Traditional brick-and-mortar retailers have been upended by the development of e-commerce giants like Amazon. The ease, variety, and competitive price provided by internet retailers have prompted existing merchants to adapt or risk obsolescence.

This disruption is not confined to retail; it spreads to numerous areas such as transportation, healthcare, and finance. In transportation, the development of ride-sharing companies like Uber and Lyft has transformed the way people travel throughout cities. These platforms employ technology to deliver a more efficient, convenient, and frequently cost-effective alternative to conventional taxis. Similarly, in healthcare, telemedicine has evolved as a feasible alternative for consultations and non-emergency medical treatment, decreasing the need for actual trips to hospitals or clinics. This revolution in how healthcare services are provided is affecting patient-doctor relationships and the healthcare sector as a whole.

Furthermore, the financial services business has undergone a huge digital upheaval. Fintech businesses have utilized technology to deliver novel financial solutions, from peer-to-peer lending platforms to digital wallets and robo-advisors. These disruptors have simplified procedures, lowered prices, and enhanced accessibility to financial services, frequently bypassing the need for established banks. As a consequence, existing financial institutions are under pressure to adapt and incorporate digital technology to stay competitive in an increasingly digital market.

Digital disruption is driven by a convergence of forces. Advances in processing power, connection, and the spread of smartphones have provided a fertile environment for digital innovation. Additionally, the accessibility of big data and the development of advanced analytics tools allow organizations to acquire significant insights into customer behavior, tastes, and market trends. This information helps firms to customize their goods and services to match the shifting expectations of the digital era.

While digital disruption brings about enormous advantages, it also raises concerns for both organizations and society at large. Incumbent enterprises that fail to adapt risk losing their market share or perhaps face extinction. Moreover, the quick speed of technology development may produce uncertainty and worry among the workforce, especially in businesses significantly touched by digital disruption. Additionally, there are worries regarding data privacy and security as digital platforms grow crucial to numerous parts of everyday life.

In conclusion, "Digital Disruption in Traditional businesses" reflects a seismic upheaval in how businesses function, driven by the combination of digital technology and novel business models. It has altered industries such as retail, transportation, healthcare, and banking, and is defined by the replacing of established methods with more efficient, convenient, and frequently cost-effective alternatives.

This disruption is backed by developments in technology, networking, and data analytics. While bringing enormous advantages, it also offers obstacles, needing adaptation and creativity for organizations to succeed in the digital age. Additionally, society must contend with concerns connected to labor transformation, data privacy, and security as we navigate this period of fast technological development.

1.4. Social and Cultural Shifts in the Digital Era

The advent of the digital age has brought in dramatic shifts in the social and cultural fabric of countries worldwide. This period, marked by the extensive use of digital technology, has transformed how humans connect, communicate, and establish their identities. One of the most visible trends is the democratization of information. With the emergence of platforms like social media, information distribution has become decentralized. Individuals are no longer passive receivers of news, but active players in the development and transmission of material. This transformation has strengthened underrepresented voices, giving birth to a more inclusive and diversified information ecosystem.

Furthermore, the digital age has brought to a rethinking of social interactions and groups. Online platforms have allowed people to interact beyond geographical borders, building virtual communities based on common interests, opinions, or experiences. This has had both good and harmful repercussions. On one side, technology has fostered a feeling of belonging and support for those who may have felt isolated in their physical communities. On the other side, it has also prompted worries about the potential for echo chambers and the propagation of disinformation within these online groups.

The digital revolution has also given birth to a new level of self-presentation and identity building. Social media platforms, in particular, have become areas for people to create and showcase their identities to the world. This has led to a complicated interaction between authenticity and performance, as people walk the delicate line between expressing their actual selves and complying to societal norms and expectations. Additionally, the availability of online anonymity has both emancipated and complicated self-expression, enabling people to explore elements of their personalities that they may not feel comfortable discussing in their offline lives.

In terms of cultural production and consumption, the digital age has transformed the way we produce, disseminate, and consume material. The barriers to entry for content production have dramatically dropped, enabling individuals and groups to develop and share their creative works on a worldwide scale. This has led to a democratization of cultural expression, disrupting conventional hierarchies within the arts and media sectors. However, it has also highlighted problems about copyright, intellectual property, and the survival of creative professions in an era of pervasive digital replication.

In conclusion, the digital age has brought about a seismic upheaval in social and cultural relations. It has democratized knowledge, altered social interactions, transformed identity building, and revolutionized cultural production and consumption. While these developments have brought about new possibilities for connection, expression, and creativity, they have also highlighted critical considerations about privacy, authenticity, and the ethical consequences of our digital relationships. As society continues to traverse this developing terrain, it is vital to critically engage with these transitions and design a digital future that is inclusive, ethical, and rewarding for everyone.

1.5. Challenges and Opportunities for the Future

"Challenges and Opportunities for the Future" is an important chapter in the book, "The Electronic Frontier: Navigating the Evolving World of Technology and Innovation in the Digital Age for a Brighter Future." This chapter looks into the complicated environment that lies ahead, giving a balanced evaluation of both the possible hurdles and enticing chances that await civilization in the digital age.

One of the key difficulties emphasized in this chapter refers to the fast speed of technological innovation. The exponential increase of innovation, especially in sectors like artificial intelligence, quantum computing, and biotechnology, raises worries about ethical and legal frameworks maintaining pace. The potential for unforeseen repercussions or abuse of these technology needs strong governance frameworks. Striking the correct balance between fostering innovation and maintaining public interests will be a tough issue for politicians and regulatory organizations.

Additionally, the chapter underlines the essential need for digital literacy and education. As technology grows increasingly incorporated into everyday life, a digital gap threatens to worsen existing socioeconomic imbalances. Without accessible and thorough instruction on digital tools and platforms, parts of the population risk being left behind. Bridging this barrier is crucial for guaranteeing inclusion and fair opportunity for everyone, regardless of financial background.

Furthermore, the chapter goes into cybersecurity as a major concern for the future. With the increased interconnection of devices and systems, the danger landscape has grown rapidly. Cyberattacks, ranging from data breaches to sophisticated state-sponsored attacks, pose major threats to people, organizations, and governments alike. Adapting to these shifting threats demands ongoing monitoring, powerful defensive systems, and worldwide collaboration in cybersecurity initiatives.

Despite these limitations, the chapter also underlines the immense potential that lay ahead. One of the most intriguing features is the potential for technology to solve critical global concerns, such as climate change and healthcare. Innovations in renewable energy, smart city planning, and precision medicine have the power to transform how we address these complicated concerns. Moreover, the digital era provides unparalleled channels for collaboration and knowledge-sharing on a worldwide scale, allowing joint efforts towards meaningful answers.

In conclusion, "Challenges and Opportunities for the Future" presents a complete assessment of the multidimensional terrain that awaits in the digital era. It highlights the necessity for proactive regulatory frameworks, underlines the relevance of digital education, and underscores the criticality of cybersecurity measures. Simultaneously, it demonstrates the great potential for technology to solve critical global concerns. Navigating this scenario will need a coordinated effort from people, corporations, governments, and the global community at large.

Chapter 2: The Internet and Its Ecosystem

2.1. Birth and Evolution of the World Wide Web

The development and growth of the World Wide Web (WWW) represents a critical milestone in human history, transforming the way information is accessed, exchanged, and disseminated. Conceived by Sir Tim Berners-Lee in the late 1980s, the WWW arose as a reaction to the rising demand for a seamless platform to transmit digital information. Berners-Lee's vision was not only about constructing a network of computers, but rather a worldwide system where papers could be connected, enabling people to traverse smoothly between them. This notion created the groundwork for a dynamic, linked network of information, converting the internet from a collection of isolated data repositories to a lively, interwoven tapestry of knowledge.

The WWW's early years were marked by a time of intense experimentation and progress. Berners-Lee's development of the Hypertext Transfer Protocol (HTTP) and the HyperText Markup Language (HTML) were significant accomplishments. HTTP allowed for the retrieval and transport of information over the internet, while HTML offered the framework for generating and structuring web pages. These standards constituted the backbone of the WWW, allowing the establishment of a user-friendly interface that anybody, regardless of technical skill, could explore.

As the 1990s progressed, the WWW underwent an incredible surge in popularity and usefulness. A notable turning point was the introduction of the Mosaic web browser in 1993, created by Marc Andreessen and Eric Bina. Mosaic introduced the notion of graphical user interfaces for navigating the web, transforming the user experience and democratizing access to online material. This breakthrough cleared the door for a larger audience to participate with the online, fueling the rise of personal websites, forums, and early forms of e-commerce.

The mid-to-late 1990s experienced a boom of business interest in the web. Companies realized the promise for the WWW as a platform for commerce, leading to the creation of internet titans like Amazon and eBay. This time also witnessed the emergence of search engines, with Google being a famous example. These engines offered users with the capacity to easily seek content among the ever-expanding vastness of the web.

The early 2000s brought another key phase in the WWW's growth with the emergence of Web 2.0. This approach emphasizes user-generated content, social networking, and collaborative platforms. Websites like Wikipedia, YouTube, and Facebook became indicative of this transition, allowing users to not only consume information but also contribute to its development. The barriers between content creators and consumers dissolved, giving birth to a participatory cyber culture.

Today, the WWW continues to grow, spurred by advances in technologies like Artificial Intelligence, Blockchain, and the Internet of Things. The spread of mobile devices and the arrival of high-speed internet have further hastened this transition, making the online a vital part of daily life. As we go ahead, the concerns of privacy, security, and digital divide remain critical, underlining the need for ethical and inclusive ways to determine the future of the World Wide Web. In this digital era, the origin and expansion of the WWW serve as a tribute to humanity's potential for creativity and collaboration on a global scale.

2.2. Internet Governance and Regulation

Internet governance and regulation have become crucial in the digital era as the global community grapples with the problems provided by the ever-expanding virtual realm. This multidimensional subject contains regulations, protocols, and frameworks that determine how the internet runs and how users interact with it. It includes a complex interaction of parties, including governments, non-governmental organizations, IT corporations, and civil society. One of the key parts of internet governance is finding a careful balance between keeping an open, free, and accessible internet while simultaneously assuring security, privacy, and compliance with regulatory frameworks.

Central to internet governance is the notion of a multistakeholder approach, which promotes inclusion and representation from many interest groups. This approach acknowledges that no one body should wield unilateral power over the internet. Instead, it pushes for cooperation between governments, civic society, business sector players, and technical specialists to cooperatively design policies and standards. This inclusive strategy aspires to stimulate innovation, defend human rights, and maintain democratic norms inside the digital sphere.

Moreover, the subject of jurisdictional borders and legal frameworks offers a substantial difficulty in internet administration. With the internet crossing geographical bounds, problems of jurisdiction become difficult. Determining which laws apply and which bodies have the ability to enforce them is a perennial challenge. This is further exacerbated by various legal systems and cultural values among nations. Striking a balance between respecting national sovereignty and guaranteeing a global, interoperable internet is a continuous effort.

Privacy and data protection are key components of internet governance. As people increasingly disclose personal information online, maintaining their privacy has become a critical issue. Various rules, such as the European Union's General Data Protection Regulation (GDPR), have been implemented to create explicit principles for data processing and protection. Balancing the demand for innovation and data-driven services with the necessity to preserve user privacy remains a major concern for regulators and lawmakers.

Cybersecurity also plays a key part in internet governance. With the increase of cyber threats and assaults, governments and businesses are obligated to adopt steps to defend their networks and data. This entails creating standards for encryption, authentication, and network security protocols. Furthermore, international collaboration is vital in countering cybercrime, since attackers typically operate across boundaries.

In conclusion, internet governance and regulation reflect a sophisticated and dynamic framework that strives to negotiate the challenges of the digital era. The multistakeholder approach, jurisdictional problems, privacy concerns, and cybersecurity considerations are all essential features of this dynamic sector. Striking a balance between maintaining a free, open, and accessible internet while simultaneously guaranteeing security and compliance with regulatory norms is a difficult endeavor that demands continual cooperation and flexibility across worldwide parties. As technology continues to evolve, internet governance will remain a vital sector in influencing the future of the digital world.

2.3. Online Communities and Social Networks

Online communities and social networks have radically revolutionized the way humans communicate, exchange information, and develop connections in the digital age. These virtual spaces serve as platforms for people with common interests, goals, or affiliations to connect, communicate, and collaborate regardless of geographic boundaries. This phenomena has had substantial ramifications for several parts of society, including social dynamics, information diffusion, and even commercial strategies.

One of the primary characteristics of online communities rests in their potential to bring together people who may not have had the chance to interact otherwise. Through common forums, interest groups, or social media platforms, individuals may connect around interests, occupations, or causes. This develops a feeling of belonging and gives an opportunity for people to seek assistance, discuss ideas, or just discover like-minded folks to engage with. For example, sites like Reddit feature a diversity of specialized communities where fans may participate in conversations on subjects ranging from unique interests to professional competence.

Furthermore, the emergence of social networks has heightened the potential for knowledge transmission and viral trends. With millions of users on platforms like Facebook, Twitter, and Instagram, a single post or message may reach a wide audience in a matter of seconds. This has altered the way news circulates, and has, in certain circumstances, challenged established media channels as key providers of information. However, this quick transmission also demands for heightened vigilance about the validity and trustworthiness of the material transmitted, since disinformation and false news may also disseminate rapidly.

On a more intimate level, social networks have transformed the nature of social interactions and relationships. Platforms like Facebook and LinkedIn have revolutionized how people maintain and develop their social networks, both personally and professionally. While they offer a handy method to remain in contact with acquaintances, they also raise problems about the genuineness of online connections and the possibility for isolating people from face-to-face encounters.

From a corporate standpoint, online communities and social networks have become crucial instruments for marketing and consumer interaction. Companies employ these channels to reach their target consumers, get feedback, and establish brand loyalty. Influencer marketing, for instance, has become a common tactic where people with big followings on social networks recommend goods or services, frequently with substantial influence on customer behavior.

However, it is crucial to realize that online communities and social networks are not without their issues. Issues of privacy, cyberbullying, and the propagation of misinformation remain significant issues. Moreover, the addictive nature of these platforms and their propensity to lead to feelings of loneliness or social isolation have generated debates about their larger societal influence.

In conclusion, online communities and social networks constitute a transformational force in the digital age, transforming how people connect, communicate, and share information. Their capacity to link disparate groups of individuals, magnify information exchange, and influence consumer behavior emphasizes their significant effect on society. While they provide several advantages, it is necessary to confront the issues they represent and manage the growing environment of digital communities with a critical and educated viewpoint.

2.4. E-commerce and the Digital Marketplace

"E-commerce and the Digital Marketplace" have altered the way companies operate and customers shop. This dynamic movement in business is defined by the purchasing and selling of products and services using electronic networks, especially the internet. The influence of e-commerce on global economies and consumer behavior is significant and diverse.

One of the most major benefits of e-commerce is its accessibility and ease. Consumers may explore and buy things from the comfort of their homes, reducing the need for physical travel and providing for 24/7 access to a large assortment of commodities. This accessibility not only appeals to convenience-seeking clients but also opens up new markets for enterprises, transcending geographical borders. Small and medium-sized firms, in particular, have exploited e-commerce platforms to compete on a worldwide scale, leveling the playing field against bigger, established corporations.

Moreover, e-commerce systems offer a rich environment for data analytics and user profiling. Through cookies, transaction history, and user activity, companies may acquire priceless insights into customer preferences and trends. This data-driven strategy enables for individualized marketing techniques, giving personalised suggestions and promotions to specific consumers. Consequently, companies may improve their product offers, pricing, and marketing activities, providing a mutually advantageous exchange of value for both customers and sellers.

However, the development of e-commerce also raises concerns, notably in terms of cybersecurity and trust. With the growing flow of sensitive information such as credit card numbers and personal data, guaranteeing safe transactions is crucial. Cybersecurity measures must be strong and up-to-date to guard against hacking, phishing, and other harmful actions. Additionally, creating trust in e-commerce platforms is vital for consumer retention. Transparency in pricing, dependable delivery, and fast customer service are essential components in developing and retaining confidence in the digital marketplace.

Another significant factor is the dynamic nature of e-commerce, driven by technology improvements. The convergence of artificial intelligence, augmented reality, and virtual reality is transforming the online buying experience. AI-powered chatbots deliver quick customer service, boosting consumer happiness. Augmented reality enables buyers to digitally try on things, bridging the gap between online and in-store purchasing experiences. These innovations not only heighten user involvement but also highlight the possibility for continual development and evolution within the digital economy.

In conclusion, e-commerce and the digital marketplace have profoundly transformed the terrain of contemporary trade. Their accessibility, data-driven skills, and capacity for innovation have brought them to the forefront of global corporate practices. Nevertheless, the concerns of cybersecurity and trust remain key considerations. As technology continues to grow, e-commerce is set to play an ever more significant role in the future of commerce, bringing both possibilities and problems for companies and consumers alike.

2.5. Internet of Things (IoT) and Its Implications

The Internet of Things (IoT) signifies a fundamental change in the way we interact with the world around us. It refers to the interconnection of common things, devices, and systems over the internet, allowing them to gather and share data. This connection offers up a range of possibilities and has far-reaching ramifications across numerous fields.

One of the most important consequences of the IoT is its potential to change businesses and sectors via greater efficiency and automation. In manufacturing, for instance, IoT-enabled sensors can monitor and improve the production process in real-time. This not only decreases operating expenses but also eliminates waste and optimizes output. Similarly, in agriculture, IoT devices may give farmers with accurate data on soil conditions, weather patterns, and crop health, leading to more sustainable and effective agricultural techniques.

Furthermore, the IoT has the potential to substantially affect healthcare and increase the quality of patient care. Wearable gadgets and remote monitoring systems may continually gather crucial health data, allowing for early diagnosis of health conditions and enabling prompt treatments. This not only enables people to take charge of their own health but also eases the pressure on healthcare systems by lowering hospital admissions and emergency department visits.

Privacy and security problems emerge as a significant consequence of the IoT. With the growth of networked devices, the quantity of personal and sensitive data being gathered rises dramatically. Ensuring strong security measures, encryption mechanisms, and consent-based data exchange policies become crucial. The possibility for illegal access or breaches might lead to catastrophic implications for people and companies alike.

Moreover, the IoT plays a crucial role in urban planning and the creation of smart cities. Through the deployment of sensors and linked infrastructure, communities may minimize energy usage, control traffic flow, and improve public services. This leads to more sustainable and livable urban areas. However, it also raises problems about data ownership, governance, and the possibility for technologically-driven societal inequities.

In conclusion, the Internet of Things is a revolutionary force with wide-ranging ramifications across numerous industries. Its potential to boost efficiency, change industries, and improve quality of life is considerable. Nevertheless, resolving concerns of privacy, security, and equal access will be important in tapping the full potential of the IoT. As society continues to incorporate these technologies into our everyday lives, intelligent analysis and proactive steps will be needed to manage this expanding digital terrain.

Chapter 3: Innovations in Information Technology

3.1. The Rise of Artificial Intelligence

"The Rise of Artificial Intelligence" highlights a significant point in the history of technology and its incorporation into daily life. This phenomena indicates a fundamental change in how computers interpret information and make judgments. At its heart, Artificial Intelligence (AI) refers to the capacity of computers to replicate human cognitive capabilities such as learning, problem-solving, and decision-making. This is done via the usage of complicated algorithms and enormous datasets, allowing AI systems to spot patterns, deduce insights, and make predictions. This transition has not only altered businesses ranging from healthcare to finance, but also created enormous social repercussions.

One of the most crucial elements of the emergence of AI is its influence on automation across numerous industries. Tasks that traditionally needed extensive human interaction are now being completed with incredible efficiency and precision by machines. In manufacturing, for instance, AI-driven robots have not only boosted output rates but have also dramatically decreased mistakes. This transition has not only enhanced productivity but has also led to a reevaluation of the role of human labor in these sectors. While certain activities are being automated, there is an increasing focus on the necessity for human abilities such as creativity, critical thinking, and emotional intelligence, which remain irreplaceable by computers.

Moreover, AI is playing an increasingly crucial role in healthcare, transforming diagnostic and therapeutic procedures. Machine learning algorithms, for example, are now capable of analyzing enormous medical information to detect patterns suggestive of illnesses. This has led to earlier and more accurate diagnosis, eventually improving patient outcomes. Additionally, AI-powered robotic surgical devices have boosted the accuracy and minimally invasive nature of procedures, leading to shorter recovery periods and lower risks. This illustrates the revolutionary potential of AI in boosting the quality and accessibility of healthcare services globally.

However, the emergence of AI also brings out a set of ethical and social issues that deserve serious examination. Issues concerning privacy, data security, and bias in algorithmic decision-making have become major considerations. The acquisition and exploitation of huge quantities of personal data to train AI models raise problems about the limits between convenience and the preservation of individual rights. Moreover, the potential for bias in AI systems, arising from the data they are trained on, underscores the need for openness, accountability, and continual monitoring in the development and deployment of these technologies.

In conclusion, "The Rise of Artificial Intelligence" represents a transformational age in the field of technology and its influence on society. The integration of AI into many businesses has enhanced efficiency, accuracy, and innovation, transforming areas from manufacturing to healthcare. Nevertheless, this progress comes with its share of ethical and social difficulties, driving the need for rigorous regulation and responsible growth. The answer lies in utilizing the potential of AI to boost human talents while ensuring that the benefits are shared fairly and the hazards are addressed. This continuous progress represents a crucial point in the trajectory of technology and its role in influencing the future.

3.2. Machine Learning and Deep Learning

Machine Learning (ML) and Deep Learning (DL) are key components of current artificial intelligence. They symbolize the progression of computer systems from rule-based programming to autonomous learning models capable of making sophisticated choices based on patterns in data.

Machine Learning provides the cornerstone of this progress. At its essence, ML is the technique of allowing computers to learn from data rather than being explicitly programmed. It depends on algorithms that spot patterns, learn from them, and make predictions or judgments without human interaction. Supervised learning, unsupervised learning, and reinforcement learning are the major paradigms of ML. Supervised learning includes training a model using labeled data, giving a clear relationship between input and output. Unsupervised learning, on the other hand, works with unlabeled input, enabling the model to discover underlying structures and patterns autonomously. Reinforcement learning presents a reward-based approach, where the model learns by performing actions and getting feedback.

Deep Learning, a subclass of ML, has attracted substantial interest owing to its extraordinary capabilities. It's defined by the usage of artificial neural networks, which replicate the structure of the human brain, made of numerous layers of linked nodes. This architecture allows DL models to handle enormous volumes of data and extract detailed characteristics, leading to very accurate predictions. Convolutional Neural Networks (CNNs) excel in image identification tasks, whereas Recurrent Neural Networks (RNNs) are good at processing sequential data. The development of specialized architectures like Generative Adversarial Networks (GANs) and Transformer models has further widened the applicability of deep learning into domains such as image synthesis and natural language processing.

The practical applications of Machine Learning and Deep Learning are broad. In healthcare, these technologies have transformed diagnosis and treatment planning by evaluating medical pictures and patient information. Moreover, in finance, ML algorithms are applied for fraud detection, risk assessment, and portfolio management. Recommender systems, as seen in streaming platforms and e-commerce, employ ML to deliver tailored content and product recommendations. Self-driving vehicles depend heavily on DL for real-time perception and decision-making, highlighting the potential for these technologies to change the future of transportation.

However, it's vital to realize the problems connected with ML and DL. The necessity for huge, high-quality datasets is a fundamental need for training effective models. Additionally, the "black box" aspect of deep learning might make it tough to comprehend and explain its judgments, posing ethical problems, especially in vital sectors like healthcare. Moreover, concerns of bias in training data and the possibility for reinforcing existing preconceptions raise substantial ethical difficulties.

In conclusion, Machine Learning and Deep Learning stand at the vanguard of the artificial intelligence revolution. Their capacity to learn from data and make intelligent conclusions has spurred breakthroughs in different sectors. Yet, ethical issues and problems connected to data quality and interpretability underline the significance of responsible development and implementation of these powerful tools. As they continue to improve, Machine Learning and Deep Learning have the possibility of driving innovation and altering our engagement with technology in the years to come.

3.3. Cloud Computing and Big Data

Cloud computing and big data are two linked technology breakthroughs that have altered the way organizations handle and analyze information. Cloud computing refers to the supply of computer services, including storage, processing, and applications, via the internet. This strategy has various advantages, such as scalability, cost-effectiveness, and accessibility from anywhere in the globe. Big data, on the other hand, applies to the huge amount of organized and unstructured data created by humans, devices, and systems. It provides particular issues in terms of storage, processing, and analysis, which cloud computing solves efficiently. This symbiotic link has become a cornerstone of contemporary information technology.

One of the primary benefits of cloud computing is its scalability. Traditional on-premise solutions frequently demand considerable upfront expenditures in hardware and infrastructure, making it harder for firms to adjust to variable workloads. Cloud services, however, enable enterprises to scale resources up or down as required, giving a flexible and cost-efficient alternative. This flexibility is especially critical when dealing with large data, since the amount of information might fluctuate dramatically over time. For example, an e-commerce platform may suffer an increase in traffic during holiday seasons, demanding more processing power and storage, which may be smoothly handled by cloud services.

Additionally, cloud computing enables better accessibility and collaboration possibilities. By putting data and apps in the cloud, workers may access crucial information from any place with an internet connection. This has altered the way firms function, allowing remote work and worldwide cooperation. For big data applications, this accessibility is vital, since it enables data scientists and analysts to work on datasets collaboratively, regardless of their physical location. It supports a more efficient and agile approach to data-driven decision-making.

Big data, with its huge and complicated datasets, would be unmanageable without the computational powers given by cloud computing. The sheer amount and diversity of data sources need significant processing capabilities that conventional on-premise systems may fail to supply. Cloud systems provide the required computing resources on-demand, allowing organizations to handle and analyze huge data effectively. This skill is important in different areas, from healthcare and finance to marketing and logistics, where data-driven insights play a crucial role in decision-making.

Furthermore, the convergence of cloud computing and big data has fostered advancements in data analytics and machine learning. Cloud providers provide a variety of tools and services intended expressly for data processing and analysis. This includes strong data warehousing systems, machine learning platforms, and data visualization tools. These services help organizations to gain significant insights from their data, enabling informed decision-making and competitive advantage. For instance, in e-commerce, big data analytics may be exploited to gather deep consumer insights, streamline supply chains, and boost marketing campaigns.

In conclusion, the synergy between cloud computing and big data has altered the way organizations manage, process, and create value from their data. The scalability and accessibility afforded by cloud computing provide a perfect setting for processing the massive volumes connected with big data. This combination helps firms to effectively handle and analyze data, promoting innovation and informed decision-making. As these technologies continue to advance, their effect on organizations and sectors throughout the world is poised to intensify, ushering in a new age of data-driven greatness.

3.4. Cybersecurity in the Digital Age

"Cybersecurity in the Digital Age" is a key feature of our increasingly linked environment. With the rise of digital technology and the availability of the internet, securing sensitive information and systems has become important. This topic comprises a diverse strategy to guard against unauthorized access, data breaches, and harmful assaults on digital infrastructure. It requires a mix of technology solutions, regulations, and user awareness. In the face of increasing threats, knowing the landscape of cybersecurity is vital for people, organizations, and governments alike.

One of the key issues in the domain of cybersecurity resides in the ever-evolving nature of cyber attacks. Malicious actors continuously alter their strategies, exploiting flaws in software, networks, and human behavior. This needs a dynamic and proactive strategy to protection, since static security solutions may rapidly become outdated. As such, cybersecurity experts must remain current of developing threats and technology, utilizing agile techniques to enhance digital defenses.

Moreover, cybersecurity is not isolated to a particular industry. It pervades practically every element of contemporary civilization, from personal gadgets to essential infrastructure. This vast breadth underlines the interdependence of our digital world, underlining that weaknesses in one sector might potentially have far-reaching implications. For instance, a compromise at a financial institution may have cascading effects on consumers' wallets, underlining the necessity for rigorous security measures at all levels of operation.

Additionally, the human factor plays a key part in cybersecurity. While technology offers a barrier against many dangers, it is not flawless. Social engineering strategies, such as phishing and pretexting, focus on abusing human psychology to obtain unwanted access. Hence, education and awareness-building programs are key components of any cybersecurity plan. Training employees to notice and react to suspicious behaviors may dramatically lower the risk of successful cyberattacks.

Furthermore, cybersecurity goes beyond the area of defense and covers incident response and recovery. In the case of a breach, having a well-defined strategy for recognizing, mitigating, and minimizing the effect is vital. This entails forensic analysis, legal compliance, and, in certain situations, public communication. A well-prepared incident response strategy may not only limit harm but also assist to recover confidence and reputation in the wake of a cyber disaster.

In conclusion, "Cybersecurity in the Digital Age" serves as a keystone in our contemporary, technology-driven society. Its relevance derives from the ubiquitous nature of digital networking and the changing threat environment. It demands a multi-layered strategy, spanning technology solutions, legislative frameworks, and user education. Moreover, a complete grasp of cybersecurity is vital for maintaining the integrity, confidentiality, and availability of digital assets. As the digital world continues to develop, so too must our plans for cybersecurity, guaranteeing a safe and resilient digital future.

3.5. Blockchain Technology and Cryptocurrencies

Blockchain Technology and Cryptocurrencies have emerged as revolutionary forces in the digital realm. The blockchain, at its foundation, is a decentralized ledger system that records transactions across a network of computers. Each block comprises a series of transactions, and if validated, it is connected to the preceding block, establishing a chain of blocks. This decentralized design assures transparency and immutability, since modifying one block would need the consensus of the whole network. This breakthrough technology has cleared the door for a number of applications outside cryptocurrencies, including supply chain management, voting systems, and even smart contracts.

Cryptocurrencies, the most well-known use of blockchain, are digital or virtual currencies that leverage cryptographic methods to safeguard and verify transactions. Bitcoin, the pioneering cryptocurrency, developed by an unidentified individual under the pseudonym Satoshi Nakamoto, heralded the beginning of this financial revolution. It offers an alternative to existing centralized financial systems by enabling peer-to-peer transactions without the need for middlemen like banks. This move not only challenges existing financial standards but also presents the prospect for financial inclusion for individuals without access to conventional banking services.

One of the primary benefits of blockchain technology is in its security aspects. The decentralized structure of the blockchain makes it immune to hacking and fraud. Transactions are validated by numerous nodes in the network, making it exceedingly impossible for any one entity to modify the data. Additionally, the cryptographic mechanisms used to safeguard the blockchain give a high degree of security against illegal access. This heightened protection has gained attention from several areas, including banking, healthcare, and supply chain management, where data accuracy and security are crucial.

However, issues do exist within the world of blockchain technology and cryptocurrencies. Scalability remains a big challenge, since the process of verifying transactions may be resource-intensive, resulting to slower transaction times as the network increases. Energy consumption is another problematic issue, especially in the case of proof-of-work based blockchains like Bitcoin, which demand substantial processing power. Additionally, regulatory frameworks are still changing, and the decentralized nature of blockchain creates significant legal and governance difficulties.

Looking forward, the potential for blockchain technology and cryptocurrencies is huge. Beyond financial applications, the technology shows potential for changing numerous areas, including healthcare, supply chain management, and voting systems. Moreover, continuing research and development are focused on tackling scalability and energy consumption challenges, with the objective of making blockchain more accessible and sustainable. As regulatory frameworks continue to grow, a careful balance must be achieved to encourage innovation while addressing concerns about legality, security, and consumer protection.

In conclusion, blockchain technology and cryptocurrencies represent a paradigm change in the way we view and interact with digital assets and information. The decentralized, secure, and transparent nature of blockchain has the potential to transform multiple sectors and reinvent old financial structures. While obstacles exist, continuous research and development, coupled with developing legal frameworks, are moving this technology towards a future of widespread acceptance and integration into our everyday lives.

Chapter 4: The Digital Economy and Entrepreneurship

4.1. Startups and the Entrepreneurial Spirit

"Startups and the Entrepreneurial Spirit" captures the vitality and creativity that define the modern corporate scene. This domain symbolizes the mentality of risk-taking, inventiveness, and an unyielding trust in transformational ideas. At its heart, a startup is not only an economic enterprise, but a representation of an entrepreneur's desire to challenge old paradigms and provide revolutionary ideas. The entrepreneurial spirit is distinguished by a distinct collection of attributes, including flexibility, resilience, and a constant pursuit of perfection. These traits help startups to handle the hurdles inherent in their birth and evolution, separating them from more traditional company models.

One of the distinguishing traits of startups is their passion for innovation. Entrepreneurs typically aim to address holes in the market or propose wholly innovative ideas. This innovation-centric strategy may lead to the creation of goods or services that disrupt sectors, providing new avenues of economic and societal influence. Startups are notorious for their capacity to question existing conventions, providing an atmosphere where innovative thinking and out-of-the-box problem-solving flourish. This culture of innovation not only pulls the business ahead but may also encourage greater industry-wide breakthroughs.

Moreover, startups are characterized by a great degree of agility. They are flexible entities, capable of turning fast in response to market input, technical advances, or unanticipated problems. This flexibility is vital in an atmosphere where change is the only constant. Unlike bigger, more established organizations, startups have the benefit of being able to execute changes swiftly, allowing them to exploit new opportunities and overcome hurdles with relative ease.

Resilience is another cornerstone of the entrepreneurial spirit. The route of a company is filled with uncertainty, setbacks, and frequently, outright failures. However, it is exactly at these times of difficulty that the real mettle of an entrepreneur is tested. The capacity to bounce from setbacks, learn from mistakes, and continue in the face of adversity is what sets successful companies distinct. This persistence is typically motivated by an unshakeable conviction in the validity of the concept and a dedication to achieve its potential significance.

Furthermore, entrepreneurs are generally propelled by a sense of purpose beyond profit. Many entrepreneurs are motivated by a desire to solve critical social or environmental concerns. This social conscience is a potent motivator, recruiting like-minded people to the startup ecosystem and linking the enterprise with a bigger cause. Startups with a clear social or environmental emphasis not only contribute to good change but also connect well with customers who increasingly respect firms that emphasize purpose alongside profit.

In conclusion, "Startups and the Entrepreneurial Spirit" constitute a dynamic force within the modern corporate environment. They embrace innovation, flexibility, resilience, and a sense of purpose that sets them different from more typical company structures. The entrepreneurial spirit motivates businesses to defy existing conventions, innovate persistently, and endure in the face of hardship. Through their creative techniques and dedication to good change, startups are not only altering sectors but also contributing to greater social advancement.

4.2. Venture Capital and Funding Trends

Venture capital plays a vital role in the innovation ecosystem, giving critical financial assistance to startups and rising firms. This sort of investment has undergone considerable adjustments and movements during the last several decades. One noticeable trend is the rising variety of venture capital sources. Traditionally dominated by institutional investors, such as venture capital companies and corporate venture arms, fresh actors have joined the picture. Angel investors, crowdsourcing sites, and even government-backed programs have become significant sources of cash for companies. This diversification has not only boosted the availability of capital but has also democratized access, enabling a larger range of entrepreneurs to seek investment.

Furthermore, there has been a substantial trend towards strategic investments. In the past, venture capital firms generally sought speedy exits via initial public offerings (IPOs) or acquisitions. However, a rising number of investors are taking a long-term approach, realizing the benefit of developing companies into viable, market-leading businesses. This tendency correlates with the larger realization of the need of sustainable development versus quick but possibly unsustainable expansion. Consequently, venture capital companies are increasingly becoming active partners, offering not just funding but also essential strategic advise and operational skills.

Additionally, there has been a spike in impact investment inside the venture capital industry. This indicates a substantial divergence from the exclusively profit-driven goals of previous investors. Impact investors strive to fund firms that show a dedication to social, environmental, or ethical concerns. This tendency shows an increasing understanding of the interdependence of corporate and society well-being. As a consequence, firms that can show a beneficial effect with financial viability typically find themselves in great demand among a new breed of socially aware investors.

Moreover, geographical diversification of venture capital activity has gained significance. While Silicon Valley has long been considered the core of venture capital, burgeoning technological clusters throughout the globe have grabbed attention. Cities like New York, London, Singapore, and Tel Aviv are increasingly competing as dynamic hubs for innovation and investment. This decentralization not only helps businesses to tap into a larger pool of possible investors but also develops varied viewpoints and approaches to entrepreneurship.

In conclusion, the landscape of venture capital and financing patterns is developing fast. Diversification of financing sources, a move towards strategic investments, the emergence of impact investing, and geographic diversification are among the important themes affecting this field. These advances reflect a more dynamic, inclusive, and socially aware approach to venture capital, eventually leading to a more strong and sustainable innovation ecosystem. As entrepreneurs continue to negotiate this dynamic terrain, adaptation and a clear awareness of these patterns will be critical for gaining the capital required to move their enterprises ahead.

4.3. Ecosystems for Innovation and Incubation

"Ecosystems for Innovation and Incubation" refers to the dynamic surroundings that stimulate creativity, experimentation, and the creation of new ideas and technologies. These ecosystems serve a vital role in growing startups and entrepreneurial enterprises, providing them with the essential resources, mentoring, and infrastructure to flourish in a competitive market. This investigation will dig into the fundamental components of innovation and incubation ecosystems, demonstrating their relevance in stimulating technical progress and economic development.

Firstly, a fundamental component of innovation ecosystems is the involvement of multiple stakeholders, including entrepreneurs, investors, academics, and governmental institutions. This variety delivers a broad range of ideas and experience, offering a fertile foundation for cooperation and knowledge sharing. Entrepreneurs generally offer new ideas and a zeal for change, while investors contribute vital cash and commercial experience. Academic institutions supply cutting-edge research and a pool of skilled employees, while governmental entities may give legislative assistance and efforts to stimulate innovation.

Furthermore, physical infrastructure and co-working spaces are important to these ecosystems. These locations offer entrepreneurs with a favorable atmosphere for brainstorming, collaboration, and experimentation. They generally contain shared resources, such as state-of-the-art laboratories, high-speed internet, and conference rooms, which may drastically cut operating expenses for early-stage enterprises. Additionally, the closeness of companies inside these venues allows spontaneous networking and information exchange, building a feeling of community that is crucial for development and learning.

Mentorship and assistance are another key part within innovation and incubation environments. Established entrepreneurs, industry experts, and seasoned professionals play crucial roles in mentoring startups. They give vital insights, share their experiences, and provide strategic guidance, helping budding enterprises overcome hurdles and make educated choices. This mentoring not only speeds the learning curve for companies but also helps them avoid frequent errors.

Access to financing sources is likely one of the most crucial components of an efficient innovation ecosystem. Startups need financial help to build prototypes, grow operations, and reach the market. This money may come from a number of sources, including venture capitalists, angel investors, government grants, and crowdsourcing sites. A well-functioning ecosystem fosters linkages between entrepreneurs and various financing sources, ensuring that bright ideas receive the financial support they need to develop and prosper.

In conclusion, "Ecosystems for Innovation and Incubation" are the core of entrepreneurial success in the digital era. By offering a collaborative, resource-rich environment, these ecosystems promote creativity, accelerate development, and drive technical innovations. The interaction of varied stakeholders, physical infrastructure, mentoring, and access to capital produces a lively and dynamic ecosystem that helps companies to translate ideas into effective technologies. Consequently, establishing and maintaining these ecosystems is important for generating economic success and technical advancement in the ever-evolving terrain of the digital age.

4.4. Digital Marketing and E-commerce Strategies

Digital marketing and e-commerce tactics have become key components of contemporary corporate operations. In an era when customers are increasingly linked to digital platforms, it is vital for companies to adapt and exploit these channels efficiently.

One major component of digital marketing is its capacity to give exact targeting. Through tools like demographic data and user behavior analysis, firms may adapt their marketing efforts to particular target groups. This ensures that resources are deployed effectively, reaching people most likely to interact with the product or service. For instance, a fashion firm may aim its adverts towards individuals who have showed interest in comparable trends, boosting the possibility of conversions.

Moreover, the real-time aspect of digital marketing provides for agility and reactivity. Unlike conventional advertising, which may need weeks to develop and deploy, digital campaigns may be altered on the go. This agility helps firms to promptly adapt to market changes, client input, or even unanticipated occurrences. For instance, a corporation may swiftly adjust its message or offers in reaction to shifts in customer preferences or incoming industry news.

In combination with digital marketing, e-commerce methods have revolutionized the way items and services are purchased and sold. The ease of internet buying has altered consumer behavior, giving a smooth and frequently customized experience. User-friendly interfaces, secure payment gateways, and trustworthy delivery services have instilled confidence in online purchases, further enhancing e-commerce's appeal.

Furthermore, e-commerce brings up new options for market development and worldwide reach. A small local firm may now get into foreign markets with remarkable ease, avoiding many of the hurdles associated with conventional exports. This globalization component of e-commerce not only broadens a business's client base but also exposes it to varied cultural and consumer tastes, which may be used for innovation and product development.

However, among the benefits, obstacles exist in the domain of digital marketing and e-commerce. The competitive environment is strong, necessitating ongoing innovation and distinction. Furthermore, problems around data privacy and security are crucial, demanding comprehensive procedures to preserve sensitive information. Additionally, keeping client trust and loyalty in the digital age is vital, given the enormous diversity of alternatives accessible to customers.

In conclusion, digital marketing and e-commerce tactics have emerged as vital instruments for organizations in the digital era. The accuracy of targeting, real-time flexibility, and worldwide reach enabled by these tactics provide great possibilities for development and market presence. Nevertheless, companies must negotiate the competitive environment and solve concerns of data security and consumer trust to fully benefit on the potential given by the digital frontier.

4.5. The Gig Economy and Future of Work

"The Gig Economy and Future of employment" signifies a fundamental paradigm change in current labor markets, typified by the emergence of short-term, freelance, and contract-based employment arrangements. This tendency has been helped by technology platforms that link people seeking job with companies or customers in need of certain skills. One of the fundamental consequences of the gig economy is its ability to alter the conventional employer-employee relationship. Unlike the usual 9-to-5 paradigm, gig labor provides people the option to choose when, where, and how much they work. This increased liberty may be encouraging for professionals attempting to juggle different commitments or striving to create a diversified portfolio of talents.

However, this flexibility comes with its own set of obstacles and uncertainty. The gig economy generally lacks the job stability, perks, and safeguards given to regular workers. Workers involved in gig agreements often do not have access to health insurance, retirement plans, or paid leave. Moreover, the gig economy may be fundamentally unpredictable, since job prospects might change dependent on market demand and individual reputation on platforms. This unpredictability may lead to financial instability and can limit individuals' capacity to prepare for the future, notably in terms of housing, healthcare, and retirement.

The gig economy also raises problems regarding workers' rights and safeguards. As many gig workers are categorized as independent contractors, they may not be eligible to minimum wage, overtime pay, or workers' compensation. This designation may leave gig workers exposed to exploitation, with little options to resolve issues. Furthermore, the absence of collective bargaining power in the gig economy might make it difficult for employees to fight for decent salaries and working conditions, possibly leading to a race to the bottom in terms of remuneration.

From an economic standpoint, the gig economy may boost production by effectively allocating resources and offering access to specialized talents on a temporary basis. It may also act as a significant source of income for folks who may not have typical work prospects owing to different situations. However, opponents claim that the gig economy exacerbates economic inequality, since higher-skilled individuals with in-demand abilities tend to do better than those engaged in low-wage, commoditized labor.

In terms of the larger future of employment, the gig economy is symptomatic of a broader transition towards a more decentralized and technology-driven labor market. As automation technologies continue to improve, some believe that some vocations may be replaced by robots, while other possibilities arise in areas that demand distinctively human skills such as creativity, emotional intelligence, and critical thinking. This transformation needs a reevaluation of education and training programs to provide people with the skills required to flourish in this developing terrain.

In conclusion, the gig economy is a transformational force in the world of work, presenting both possibilities and problems for people, organizations, and regulators. While technology provides employees greater freedom and access to different options, it also creates economic and societal risks. Striking a balance between exploiting the advantages of the gig economy and ensuring the rights and well-being of employees will be a vital job for creating the future of employment.

Chapter 5: Privacy and Data Protection

5.1. Data Privacy in the Digital Age

In the present context of the Digital Age, the problem of data privacy has arisen as a key concern for people, corporations, and governments alike. With the expansion of digital technology, the gathering, storage, and exploitation of personal information has reached new levels. This has led to a fundamental change in the way cultures view and handle privacy.

One of the fundamental elements of data privacy in the Digital Age is the enormous amount and variety of data being created and handled. From social media interactions and online purchases to health records and location data, the quantity of information that is gathered about people is startling. This plethora of data acts as a double-edged sword, giving enormous potential for insights, creativity, and customisation, but also presenting major hazards if not managed correctly.

Furthermore, the idea of permission has been a prominent topic in arguments concerning data privacy. In an age of complicated terms of service agreements and sometimes opaque data collecting techniques, consumers are routinely asked to express permission without a complete knowledge of the repercussions. This raises crucial concerns regarding the exact degree to which people have agency over their personal information, and whether present legal and ethical frameworks are strong enough to preserve their rights.

Another crucial component of data privacy is to the involvement of technology businesses and other data-driven organizations. Tech firms, in particular, have come under investigation for their data processing methods, raising disputes over their responsibilities in ensuring user privacy. As these corporations collect vast banks of user information, worries about monopolistic power and possible abuse of data have been fundamental in the conversation around digital privacy.

Legislation and regulation play a significant role in establishing the parameters of data privacy. Governments throughout the globe are wrestling with the necessity to establish laws that strike a balance between promoting innovation and preserving the privacy rights of individuals. Notable examples include the European Union's General Data Protection Regulation (GDPR) and comparable measures in other areas. However, the dynamic nature of technology creates a continuing challenge for regulators, as they attempt to stay pace with fast developments in data processing capacity.

In conclusion, data privacy in the Digital Age is a complicated problem that touches on technical, ethical, legal, and sociological factors. The sheer amount of data created, along with the difficulties surrounding permission and the involvement of digital corporations, highlights the importance of resolving this problem. Striking a balance between innovation and privacy protection is crucial for establishing trust in the digital economy. As societies struggle with the ramifications of the data-driven age, continuing conversations and strong regulatory frameworks will be vital in defining the bounds of privacy in the years to come.

5.2. The Role of Data in Business and Government

"The Role of Data in Business and Government" is a key issue in the contemporary day, as both sectors increasingly depend on data-driven decision-making to accomplish their aims. In company, data has transitioned from being a supplemental asset to being the lifeblood of operations. Companies currently acquire large volumes of information, ranging from client preferences to market trends, allowing them to adjust goods and services to match growing wants. Through advanced analytics and machine learning algorithms, organizations can extract useful insights from this data, giving them with a competitive advantage in an ever-changing market scenario.

Furthermore, data plays a key role in boosting consumer experiences and encouraging brand loyalty. By studying client behavior, firms may customize interactions, give personalized suggestions, and provide targeted promotions. This individualized strategy not only promotes client pleasure but also generates revenue growth. It helps organizations to anticipate consumer wants, eventually leading to improved retention rates and increased client lifetime value.

In government, the importance of data is as deep. Data-driven governance is revolutionizing public sector operations, enabling for more efficient resource allocation and policy formation. For instance, via data analysis, governments may identify locations with special requirements, allowing them to distribute resources appropriately. This guarantees that public services are supplied where they are most needed, encouraging justice and equality throughout the population.

Moreover, data-driven decision-making in government may lead to better openness and accountability. Access to reliable and timely information helps individuals to monitor government operations, encouraging confidence in public institutions. Additionally, data may be used to analyze the success of policies and initiatives, allowing for modifications and improvements over time. This evidence-based approach guarantees that taxpayer monies are distributed efficiently and effectively, maximizing the delivery of public services.

However, the rising dependence on data also presents serious ethical and privacy problems. Businesses and governments must carefully weigh the advantages of data-driven decision-making with the preservation of individual privacy rights. It is vital that effective data governance structures be put in place to preserve sensitive information and guarantee compliance with legal and regulatory obligations.

In conclusion, the role of data in business and government is revolutionary, transforming the way choices are made and services are provided. Through sophisticated analytics and technology capabilities, firms may gain a competitive advantage and improve consumer experiences. In the public sector, data-driven governance leads to more efficient resource allocation, more openness, and enhanced policy efficacy. Nevertheless, it is vital to approach data utilization with ethical concerns, ensuring that privacy rights are maintained. Striking a balance between leveraging the potential of data and respecting individual privacy is a vital problem for both industries in the digital age.

5.3. GDPR and Global Data Protection Regulations

The General Data Protection Regulation (GDPR) stands as a major piece of legislation in the domain of data protection, substantially changing how companies manage personal information. Enacted by the European Union in 2018, GDPR has worldwide consequences, impacting any organization that collects or keeps data relating to EU people. This rule intends to empower people with more control over their personal data while setting rigorous duties on corporations and organizations in securing this information.

One of the essential components of GDPR is its focus on informed consent. Individuals must now be given with clear and readily accessible explanations of how their data will be used. This openness ensures that individuals are aware of the reason for which their information is being gathered, therefore allowing them to make educated choices about providing their personal data. This degree of openness not only bolsters privacy rights but also develops a foundation of confidence between enterprises and their customers or users.

Moreover, GDPR provides the "right to be forgotten," a notion that allows people the capacity to seek the erasure of their personal data under specific conditions. This option is a big leap towards digital autonomy, since it allows people to detach themselves from information that may no longer be relevant or required. It also corresponds with the developing idea of privacy in an age when digital traces may be far-reaching and long-lasting.

GDPR is not without its issues, though. Compliance takes substantial resources, especially for smaller enterprises or startups. The development of effective data security procedures and the recruitment of Data security Officers may be expensive. This has prompted some to claim that GDPR may unwittingly hamper innovation, particularly for new firms that may struggle to deploy resources to fulfill the onerous criteria.

The extraterritorial reach of GDPR has stimulated a worldwide trend towards more strong data protection regulations. Many nations and areas, such as California with the California Consumer Privacy Act (CCPA), have passed or are contemplating similar laws. This worldwide trend reflects an increasing realization of the necessity of data security and privacy in the digital era. However, the development of regional rules also provides a problem for international firms, as they must traverse a complicated web of compliance obligations.

In conclusion, GDPR stands as a watershed event in the growth of data protection, creating a new norm for privacy rights in the digital age. It positions people at the forefront of data control, demanding openness, informed consent, and the right to be forgotten. While the rule offers obstacles, especially for smaller enterprises, its global significance is evident, propelling a larger trend towards greater data security globally. Striking a balance between privacy and innovation, GDPR offers as a blueprint for future legislation in an increasingly linked and data-driven society.

5.4. Personal Privacy in a Connected World

"Personal Privacy in a Connected World" has become a key worry in today's digitally-driven world. The introduction of modern technology and the proliferation of networked gadgets have resulted to an unparalleled amount of access to personal information. This issue has sparked a critical assessment of the borders between public and private worlds, as people battle with the problems of preserving their sensitive data.

One of the core concerns surrounding personal privacy in a connected society is the enormous amount of data created and gathered. From social media interactions to online purchases, people leave a digital trail that, when collected, gives a detailed picture of their activities, interests, and even emotions. This raises problems regarding consent and awareness, since many people may not completely grasp the degree to which their personal information is being captured and exploited by numerous groups.

Moreover, the growth of monitoring technology and the ubiquity of smart gadgets have further weakened conventional concepts of privacy. Smart home systems, for instance, might unwittingly disclose sensitive information of an individual's everyday life, from their routines to their interpersonal relationships. As these gadgets grow increasingly interwoven into our lives, the possibility for unintended data disclosure rises, highlighting the need for effective privacy measures.

Legislation and regulation play a significant role in seeking to establish a balance between innovation and personal privacy. Concepts such as the General Data Protection Regulation (GDPR) in the European Union and similar projects globally attempt to empower people with more control over their personal data. These policies enforce openness and require express permission for data gathering, allowing people the agency to determine how their information is used.

However, despite these attempts, difficulties continue. The fast speed of technology progress sometimes outpaces the capacity of regulatory frameworks to keep up. This creates a dynamic where people are left open to new privacy threats, needing a continual evaluation of legal measures. Additionally, the global nature of the internet complicates problems, as jurisdictional boundaries become blurred, and enforcement of privacy rights across countries becomes a hard task.

In conclusion, "Personal Privacy in a Connected World" is a key frontier in the digital era. The massive volumes of data created, combined with the presence of networked devices, underline the need of preserving individual privacy. While law and regulation play a critical role, a comprehensive strategy is essential, incorporating education, technology innovation, and continual inspection of privacy practices. Balancing the advantages of a connected world with the requirement of preserving personal information is a complicated problem that will determine the future of digital society.

5.5. Ethical Considerations in Data Usage

"Ethical Considerations in Data Usage" is an important component of our increasingly digital environment. As we negotiate this terrain, it's vital to assess the ethical implications of how data is gathered, processed, and used.

Firstly, privacy sits at the forefront of ethical considerations in data utilization. Individuals have a basic right to manage their personal information. Unauthorized or non-consensual data acquisition breaches upon this right, possibly leading to a breach of trust. Therefore, gaining informed permission and establishing openness about data practices are crucial. Additionally, preserving sensitive data from breaches and cyber-attacks is vital in sustaining people' confidence.

Furthermore, the idea of fairness in data utilization must be addressed. It's crucial to eliminate prejudice or bias in algorithms and decision-making processes that depend on data. This guarantees that persons are treated fairly, regardless of their origin or qualities. Striving for justice demands continual monitoring and modification of algorithms, and fostering diversity and inclusion in data collecting.

Additionally, there is a duty to assess the intended purpose of data utilization. Misuse or distortion of data may have far-reaching effects. For instance, in healthcare, erroneous or misread data might lead to wrong diagnoses or treatment plans. Thus, ensuring that data is used for its intended purpose, and not repurposed in a way that may hurt persons, is a significant ethical matter.

Moreover, data management is an ethical obligation. Those charged with collecting and managing data carry the burden of guaranteeing its integrity, security, and proper usage. This applies to both persons and organizations. Implementing effective data governance policies, including encryption, access limits, and frequent audits, helps sustain this concept.

Finally, accountability is crucial in ethical data utilization. Organizations and people must accept responsibility for their actions and choices based on data. This involves admitting and addressing any errors or prejudices that may develop. Additionally, clear lines of responsibility guarantee that people have redress in case of abuse or mismanagement of their data.

In conclusion, ethical issues in data utilization are crucial in our digital age. These factors cover privacy, justice, purpose, stewardship, and responsibility. Upholding these values guarantees that data utilization benefits society as a whole, while reducing any disadvantages. Striking a balance between leveraging the power of data and protecting human rights and social values is vital for a responsible and ethical data economy.

Chapter 6: Artificial Intelligence and Society

6.1. AI in Healthcare and Medicine

AI (Artificial Intelligence) in Healthcare and Medicine represents a paradigm change in the way healthcare is given and administered. This combination of modern technology with medical research has enormous potential for improving patient outcomes, simplifying procedures, and increasing the skills of healthcare workers.

One of the key benefits of AI in healthcare resides in its potential to evaluate enormous volumes of data with speed and accuracy. Machine learning algorithms can evaluate electronic health data, medical imaging, and even genetic information to discover trends, make diagnoses, and prescribe treatment strategies. This feature has the potential to greatly minimize diagnostic mistakes, a vital part of patient safety.

Furthermore, AI-enabled predictive analytics play a key role in preventative care. By examining a patient's past data and risk variables, AI may assist healthcare practitioners identify those at high risk for specific illnesses. This enables for preventative treatments and tailored healthcare strategies, eventually leading to a decrease in the burden of chronic diseases and related healthcare expenses.

In the realm of medical imaging, AI has showed tremendous potential. Computer vision algorithms, for example, can assess radiological images such as X-rays, MRIs, and CT scans with a degree of accuracy that approaches, and in some instances exceeds, human expertise. This not only expedites the diagnosis procedure but also guarantees that potentially severe illnesses are recognized at an early, more curable stage.

Another transformational feature of AI in healthcare is its ability to better treatment planning and medication development. Through complex algorithms, AI can simulate the impacts of different treatment alternatives, supporting physicians in creating optimum care methods. Additionally, in pharmaceutical research, AI can scan enormous databases to uncover prospective drug candidates and improve the drug development process, thereby speeding the availability of life-saving therapies.

Despite its great promise, the integration of AI in healthcare is not without hurdles. Ethical issues, data privacy, and security concerns are crucial. Ensuring that patient data is utilized ethically and securely is vital to sustaining public confidence in these technologies. Additionally, there is a need for continual training and education for healthcare workers to successfully harness and comprehend AI-driven insights.

In conclusion, AI in healthcare and medicine shows significant potential in transforming the way healthcare is given. Its capacity to handle and analyze enormous volumes of data with speed and precision may considerably enhance diagnosis accuracy and treatment planning. Additionally, AI's position in preventative care and medication research has the potential to bring in a new age of customized medicine. However, it is vital that ethical, privacy, and security aspects be thoroughly addressed to fully exploit the potential advantages of AI in the healthcare industry.

6.2. AI in Education and Learning

Artificial Intelligence (AI) has emerged as a disruptive force across multiple industries, and one of the arenas where its potential is most significant is in education and learning. This integration of AI technology has the power to change conventional teaching approaches and tailor the learning experience for pupils of all ages and backgrounds. One of the key benefits comes in its capacity to modify material delivery to suit various learning styles and paces, offering a more inclusive and successful educational environment.

AI-driven educational technologies have the power to assess huge volumes of data on a student's learning habits, strengths, and shortcomings. By applying machine learning algorithms, these systems may develop individualized learning plans, adapting the curriculum to fit the particular requirements of each learner. This degree of customisation not only creates a better grasp of the subject matter but also stops learners from feeling overwhelmed or under-stimulated, eventually leading to greater engagement and retention.

Furthermore, AI's aptitude for natural language processing and comprehension allows advanced interactions between students and educational materials. This is notably visible in the creation of intelligent virtual teachers or chatbots. These technologies may offer quick feedback, answer inquiries, and lead students through challenging ideas, essentially imitating the job of a human teacher. This real-time help is vital, particularly in instances when instant explanation is crucial for understanding and growth.

Another key component of AI in education is its ability to overcome gaps in accessibility and inclusion. By merging voice recognition and synthesis technology, AI applications may aid students with impairments in overcoming challenges to conventional learning techniques. For instance, visually handicapped students may leverage AI-driven text-to-speech programs to access written information, providing a more inclusive educational environment.

However, although the advantages of AI in education are enormous, ethical problems and privacy concerns must not be neglected. The gathering and analysis of student data need strong protections to preserve the confidentiality and security of sensitive information. Additionally, there is a need for openness in how AI systems make judgments to eliminate biases and to retain responsibility in the learning process.

In conclusion, the integration of AI in education shows significant potential for the future of learning. Through individualized learning experiences, quick feedback, and enhanced accessibility, AI has the potential to change education at all levels. However, it is vital that as AI gets more completely incorporated into educational institutions, careful attention be given to ethical issues and privacy concerns, ensuring that the advantages are maximized while any threats are addressed. By doing so, the education industry can leverage the full potential of AI to create a more inclusive, engaging, and successful learning environment for all students.

6.3. AI in Autonomous Vehicles and Transportation

AI in Autonomous Vehicles and Transportation has altered the way we envisage and interact with transportation systems. This technology, at its heart, tries to connect artificial intelligence algorithms with the mechanics of vehicles to provide self-driving capabilities. The importance of this invention cannot be emphasized, as it offers a myriad of advantages ranging from greater safety to enhanced efficiency in our transportation networks.

One of the key benefits of adopting AI in autonomous cars is in the ability to substantially decrease accidents and deaths on the road. Traditional accidents are generally ascribed to human error, which autonomous cars strive to avoid by relying on sensors, cameras, and computer algorithms to make quick, real-time choices. These systems are meant to sense their surroundings and manoeuvre through complicated traffic conditions with a degree of accuracy and consistency that transcends human capabilities. Consequently, this technology has the prospect of significantly altering our approach to traffic safety.

Moreover, AI-driven autonomous cars have the potential to dramatically increase the efficiency of transportation networks. Through real-time data processing, these cars can optimize routes, predict traffic patterns, and alter speeds to reduce congestion. This not only leads to decreased travel times for people but also results in a more sustainable and eco-friendly transportation system. By making more effective use of highways, autonomous cars have the potential to lessen the environmental impact associated with transportation.

However, the deployment of AI in autonomous cars also brings its own set of obstacles. One significant worry centers on the ethical considerations of programming algorithms to make life-or-death judgments in emergency circumstances. For instance, establishing how a vehicle should prioritize the safety of its passengers vs pedestrians is a complicated moral issue that involves careful thinking and creative design. Striking a balance between safety, fairness, and ethical decision-making is a continuing problem in the development of autonomous cars.

Furthermore, there are considerable infrastructural and legal changes that must precede the broad implementation of autonomous cars. Roads and traffic systems need to be fitted with the required technology to permit communication between cars and with the larger transportation network. Additionally, a full legal structure must be constructed to handle responsibility, accountability, and compliance with safety requirements. These factors underline the need of a comprehensive strategy to incorporating AI into transportation systems.

In conclusion, the inclusion of AI in autonomous cars and transportation constitutes a key milestone in technical growth. The potential to increase safety, efficiency, and sustainability in our transportation networks is considerable. Nevertheless, it is vital to approach this shift with a careful understanding of the ethical, legal, and infrastructural problems that lie ahead. Through careful planning and cooperation between technology developers, politicians, and the wider public, we can strive towards a future where autonomous cars play a major role in designing a safer, more efficient, and sustainable transportation scene.

6.4. Ethical and Social Implications of AI

Artificial Intelligence (AI) has quickly evolved in recent years, bringing a plethora of ethical and societal consequences that deserve serious examination. One of the key concerns focuses on problems of prejudice and fairness. AI systems learn from enormous databases, and if these sources include biases, the AI may perpetuate and even worsen them. For example, in domains like employment or criminal justice, biased algorithms might result in discriminatory results, disproportionately hurting particular populations. Therefore, it is vital to build effective procedures for auditing and eliminating biases in AI systems.

Another big problem lies in the field of privacy and monitoring. As AI systems become more interwoven into our everyday lives, they create and handle huge volumes of data. This data may contain sensitive information about persons, raising problems regarding permission, data ownership, and the potential for abuse. Striking the correct balance between using data for increasing AI skills and respecting human privacy is a complicated task that needs defined legal frameworks and technical protections.

Furthermore, the influence of AI on employment and labor markets is a matter of significant dispute. While AI has the ability to boost productivity and generate new employment possibilities, it also has the power to automate some processes, possibly leading to job displacement. This needs proactive initiatives for reskilling and upskilling the workforce, as well as regulations that assure an equitable transition for people impacted by technological advancements.

Ethical concerns in AI extend to questions of responsibility and transparency. As AI systems get increasingly complicated and autonomous, it might be tough to grasp the decision-making processes underlying their activities. This "black box" dilemma raises problems regarding who should be held accountable in circumstances of AI-driven failures or damage. Establishing clear lines of responsibility and establishing tools to understand AI choices are essential steps towards ensuring that AI systems be employed responsibly and ethically.

Moreover, there are important consequences for autonomy and agency. In circumstances where AI systems make choices that directly touch persons, issues emerge concerning the degree of control and influence that people should have over these systems. Striking a balance between improving human capacities and retaining individual agency is crucial to preventing unnecessary concentration of power and ensuring that AI supports, rather than threatens, human well-being.

In conclusion, the ethical and societal consequences of AI are numerous and complicated. They touch on concerns of prejudice, privacy, employment, accountability, and autonomy. Addressing these challenges demands a multidisciplinary approach, engaging not just engineers and legislators but also ethicists, sociologists, and other stakeholders. It is vital that society collaboratively grapples with these difficulties to guarantee that AI technologies are created and used in a way that supports human values and promotes the greater good.

6.5. AI and Employment Disruption

The development of Artificial Intelligence (AI) has ushered in a new age of technical growth, but it has also brought about huge upheavals in the work environment. This disruption originates from the unique capabilities of AI systems, which can analyze large volumes of data, learn from patterns, and execute jobs with growing autonomy. As AI technologies continue to improve, industries across the board are experiencing transformations in labor relations. Traditional employment that require repetitive operations or regular decision-making processes are especially vulnerable to automation.

One of the primary consequences of AI on work is the displacement of some regular, physical, and even cognitive activities. Industries that mainly depend on manual labor, such as manufacturing and assembly-line production, have experienced a drop in the necessity for human workers as AI-powered robots and computers take over these monotonous jobs. Furthermore, cognitive functions that were traditionally the domain of experienced professionals, like data analysis or some areas of customer service, are becoming automated, possibly diminishing the necessity for human labor in these jobs.

However, it's vital to note that although AI might lead to employment displacement in some industries, it also provides prospects for job creation and change. This is notably visible in the rise of totally new job categories, such as AI developers, data scientists, and machine learning engineers. These jobs are vital in the creation, deployment, and maintenance of AI systems. Additionally, the necessity for people to work alongside AI, giving supervision, making strategic choices, and ensuring ethical issues are addressed, is becoming increasingly crucial. Thus, AI is altering the nature of labor, rather than obviating it totally.

Another component of the job disruption created by AI is the urge for upskilling and reskilling of the workforce. As AI systems take over regular jobs, there is a rising need for individuals who possess talents that complement and interact successfully with AI. This includes talents linked to data processing, critical thinking, creativity, and emotional intelligence - areas where humans thrive and where AI typically falls short. Educational institutions, politicians, and companies themselves have a critical role in enabling this transition by offering training and development programs that equip people with the skills essential to survive in the era of AI.

While there are worries about the possible job losses due to AI, it's crucial to remember that these disruptions are not consistent across all businesses or areas. Some industries, especially those that need nuanced decision-making, emotional intelligence, or human touch, are less vulnerable to automation. Furthermore, AI may boost productivity and provide efficiencies that can contribute to economic development, perhaps offsetting some of the first disruptions.

In conclusion, the integration of AI into the workforce is certainly altering the job environment. It is leading to the automation of ordinary jobs and the introduction of new job categories, demanding a change in the skills and duties of the workforce. While there are problems connected with this disruption, with careful planning, education, and policy implementation, society may harness the promise of AI to build a more productive, inventive, and inclusive workforce for the future.

Chapter 7: Cybersecurity and Digital Threats

7.1. Types of Cybersecurity Threats

"Types of Cybersecurity Threats" encompass a wide range of malicious activities that target computer systems, networks, and the sensitive data they hold. In today's interconnected digital world, understanding and addressing these threats are essential for individuals, organizations, and governments. This analysis will explore some of the most common types of cybersecurity threats and their implications.

1. Malware Attacks: Malware, short for malicious software, is a broad category of threats that includes viruses, Trojans, worms, ransomware, and spyware. These programs are designed to infiltrate systems, steal data, disrupt operations, or encrypt files for ransom. Malware attacks are prevalent and can have severe consequences, including data breaches and financial losses.

2. Phishing and Social Engineering: Phishing attacks involve tricking individuals into revealing sensitive information like passwords, credit card numbers, or

personal details. Social engineering tactics often accompany phishing, manipulating human psychology to gain access to systems. These attacks are typically initiated through deceptive emails, websites, or phone calls.

3. DDoS Attacks: Distributed Denial of Service (DDoS) attacks overwhelm a system with traffic from multiple sources, rendering it inaccessible to users. These attacks can disrupt online services, leading to financial losses and reputational damage. Hackers may use botnets to orchestrate massive DDoS attacks.

4. Insider Threats: Insider threats come from individuals within an organization who misuse their access privileges. These threats can be intentional (e.g., disgruntled employees) or unintentional (e.g., employees falling victim to phishing scams). Insider threats can result in data leaks, data theft, or sabotage.

5. Zero-Day Exploits: A zero-day exploit targets vulnerabilities in software or hardware that are unknown to the vendor or have not been patched yet. Cybercriminals capitalize on these vulnerabilities before they are discovered and fixed. These exploits

can lead to data breaches, system compromise, and significant damage.

6. Advanced Persistent Threats (APTs): APTs are sophisticated, targeted attacks often associated with nation-states or well-funded organizations. They involve a long-term, stealthy approach to infiltrate and control a target network, potentially for espionage or data theft. Detecting and mitigating APTs is challenging due to their complex and persistent nature.

7. Insider Trading: This type of threat is particularly relevant to the financial industry. Cybercriminals can use stolen insider information for illegal trading activities. Insider trading can have severe legal and financial consequences for both individuals and companies.

8. Supply Chain Attacks: Supply chain attacks occur when cybercriminals infiltrate an organization through its third-party vendors or service providers. These attacks can compromise software updates, hardware components, or other assets, potentially leading to widespread breaches.

In summary, the landscape of cybersecurity threats is dynamic and constantly evolving. As technology advances, cybercriminals find new ways to exploit vulnerabilities and target individuals and organizations. Effective cybersecurity strategies require a combination of technical defenses, user education, and proactive monitoring to detect and respond to these threats. Moreover, collaboration among governments, industries, and individuals is crucial to stay ahead of the ever-changing threat landscape and protect digital assets.

7.2. Protecting Personal Data and Identity

In an age dominated by digital technology and networked platforms, preserving personal data and identity has arisen as a key issue. The omnipresent nature of the internet and the massive troves of data collected and exchanged everyday have created fertile ground for possible breaches and exploitation. Consequently, people and organizations alike must implement stringent steps to secure sensitive information from illegal access and harmful intent.

One of the fundamentals of securing personal data is encryption technology. Encryption utilizes complicated algorithms to jumble data, leaving it unintelligible to anybody without the decryption key. This technology operates as a solid barrier against unwanted access, whether in transit or at rest. Advanced encryption algorithms are now a typical part of communication channels, ensuring that information sent across networks stays safe. Moreover, the spread of secure socket layers (SSL) and transport layer security (TLS) protocols has bolstered online transactions, assuring that financial and personal information is guarded from prying eyes.

In concert with encryption, the deployment of tight access controls is vital in preserving personal data. This entails setting up levels of authentication, so users must submit various kinds of verification before having access to sensitive information. Multi-factor authentication (MFA) has developed as a strong technique in this respect, demanding the submission of something known (like a password), something held (such as a smartphone), and something biometric (like a fingerprint). This multi-layered technique considerably decreases the risk of unwanted access, even in the case of a leaked password.

However, the success of preventive measures is dependant on the understanding and education of people on recommended practices for data security. Cybersecurity training and awareness initiatives play a crucial role in reinforcing the human aspect against social engineering approaches, such as phishing attempts. By training personnel with the expertise to recognize suspicious communications and to identify possible threats, businesses may greatly minimize the risk of successful intrusions.

Moreover, the legal environment around data protection has experienced tremendous change. Legislation such as the General Data Protection Regulation (GDPR) in the European Union and the California Consumer Privacy Act (CCPA) in the United States have set rigorous regulations on enterprises regarding the acquisition, storage, and use of personal data. These policies not only empower people with more control over their information but also place significant fines on companies failing to meet these requirements. Compliance with such restrictions has become non-negotiable for enterprises working in the digital sphere.

In conclusion, preserving personal data and identification is a priority in today's digitally-driven society. Encryption technology, effective access controls, cybersecurity knowledge, and legal compliance constitute the core of a complete data protection plan. As the digital ecosystem continues to change, so too must the measures adopted to fight against the ever-present danger of illegal access and data breaches. Only via a multi-faceted strategy can people and organizations guarantee that sensitive information stays safe in the face of an increasingly complex threat environment.

7.3. National Security and Cyber Warfare

In the fast shifting terrain of national security, the rise of cyber warfare has ushered in a new age of strategic difficulties. Unlike traditional forms of conflict, cyber warfare occurs in the virtual environment, where the battlefield is constituted of linked networks, systems, and information. This paradigm change has caused governments to reevaluate old conceptions of defense and security, realizing that the vulnerabilities of the digital realm are as essential to national interests as physical boundaries.

One of the fundamental elements of cyber warfare resides in its asymmetry. Unlike conventional warfare, when a nation's military force is frequently a main factor of success, cyber warfare provides a level playing field for both state and non-state players. A very small number of experienced hackers may possibly inflict tremendous harm on a technologically sophisticated country. This democratization of power in the digital arena has sparked a reevaluation of traditional military plans and doctrines.

Furthermore, the interconnection of global networks implies that national security is unavoidably linked with international relations and diplomacy. Cyber assaults have the potential to destroy not just a nation's essential infrastructure but also its economic stability and social fabric. Thus, the boundary between national and international security has become more blurred, demanding a more coordinated approach to cybersecurity. International standards and agreements have been suggested to set criteria for responsible conduct in cyberspace, reflecting the understanding that unilateral acts may have far-reaching implications in the linked globe.

Cyber warfare also brings a distinct set of issues in attribution. Unlike conventional combat, when identifying the aggressor is typically clear, the anonymity given by the digital environment may complicate the process of establishing guilt for a cyber strike. This ambiguity may lead to confusion and slowness in reaction, as governments wrestle with the complexity of appropriately attributing responsibility. As a consequence, the development of effective cybersecurity capabilities and information collection has become vital in deterring future aggressors.

In addition to state-sponsored cyber warfare, the emergence of non-state actors, including hacktivist groups and cybercriminal organizations, has further complicated the picture of national security. These organizations typically act outside the bounds of conventional geopolitical disputes, propelled by ideological, financial, or political motivations. Their quickness, combined with the anonymity given by cyberspace, provides a severe challenge to national security agencies entrusted with securing essential infrastructure and sensitive information.

In conclusion, the rise of cyber warfare has radically changed the dimensions of national security. The virtual war, distinguished by its asymmetry, interconnection, and issues of attribution, needs a reevaluation of established security paradigms. To traverse this digital terrain successfully, states must take a multidimensional strategy that involves not just military capabilities but also international collaboration, cybersecurity legislation, and adaptive methods to target both state and non-state actors. In this age of cyber warfare, preserving national interests needs a complete grasp of the intricate interaction between technology, geopolitics, and security.

7.4. The Role of Ethical Hacking

Ethical hacking, often known as penetration testing or white-hat hacking, comprises a significant component in contemporary cybersecurity. It includes simulating cyber-attacks on computer systems, networks, or applications with the sole objective of identifying vulnerabilities and flaws. The ethical hacker, typically recruited by corporations or governments, acts within legal and ethical bounds, aiming to uncover and repair any security gaps before bad actors may exploit them. This proactive approach to security is crucial in securing sensitive data, ensuring the integrity of systems, and sustaining confidence in the digital domain.

One of the key goals of ethical hacking is to serve as a preemptive strike against cyber threats. In an age defined by an ever-evolving environment of sophisticated hacking tactics and cyber-attacks, enterprises can ill-afford to stay inactive in their security measures. By using ethical hackers to find weaknesses, businesses may successfully inoculate themselves against prospective attacks. This not only prevents companies from any financial damages arising from data breaches but also defends their reputation and client confidence. Ethical hacking, in this perspective, acts as a critical insurance policy against the potentially disastrous results of a cyber-attack.

Moreover, ethical hacking plays a key function in compliance and regulatory adherence. In numerous businesses, especially those dealing with sensitive information such as banking, healthcare, and government, tight standards control data security and privacy. Ethical hacking offers an organized and controlled environment to examine the organization's adherence to these standards. By finding and fixing any departures from compliance norms, ethical hacking guarantees that firms satisfy their legal commitments, safeguarding them from any legal ramifications and financial fines.

Ethical hacking also adds to the continual enhancement of cybersecurity systems. The insights acquired from ethical hacking activities offer firms with important data on upcoming risks and weaknesses. This information, when utilized correctly, allows the creation of more strong security protocols and the deployment of sophisticated protection measures. Additionally, ethical hacking encourages a culture of continual development in cybersecurity understanding and procedures inside a firm. It supports a proactive posture towards security, emphasizing vigilance and adaptation in the face of increasing cyber threats.

Furthermore, ethical hacking functions as a crucial instructional tool in the field of cybersecurity. It gives a chance for security professionals to strengthen their abilities and expertise by immersing themselves in real-world circumstances. Ethical hackers typically adopt cutting-edge tactics and technologies, helping them to remain ahead of possible attackers. This skill, obtained via ethical hacking, is vital in teaching the next generation of cybersecurity specialists and ensuring that they are well-equipped to handle the challenges of an increasingly digital world.

In conclusion, ethical hacking serves as a critical pillar in current cybersecurity efforts. By proactively finding vulnerabilities, guaranteeing regulatory compliance, promoting continual improvement, and acting as an instructional tool, ethical hacking contributes greatly to the overall security posture of enterprises and governments alike. In an age when cyber threats continue to advance in complexity and size, ethical hacking is a vital technique in preserving the digital environment.

7.5. Preparing for Future Cyber Threats

"Preparing for Future Cyber Threats" is an important chapter in the book "The Electronic Frontier: Navigating the Evolving World of Technology and Innovation in the Digital Age for a Brighter Future." This chapter dives into the importance of defending against rising cyber risks, a problem that has increased in unison with the fast spread of digital technology.

First and foremost, this chapter underlines the dynamic and growing nature of cyber threats. It understands that the environment of cyber threats is in a perpetual state of change, as hackers and bad actors modify their strategies in response to developments in security measures. By noting this flexibility, the chapter invites readers to take a proactive and forward-thinking approach to cybersecurity.

Furthermore, the chapter underlines the requirement of a multi-layered defensive system. It cautions against depending exclusively on conventional security measures and instead argues for a complete strategy that comprises powerful firewalls, encryption methods, intrusion detection systems, and frequent security audits. This multimodal strategy is critical in mitigating the varied spectrum of attacks that might target various components of digital infrastructure.

Moreover, the chapter underlines the necessity of threat intelligence and information exchange. It highlights that organizations, both public and commercial, need to build processes for acquiring and distributing information on new cyber dangers. This collaborative strategy offers a collective defense against cyber threats, ensuring that information and insights are shared across sectors, so boosting the overall security posture.

The chapter also pays attention to the need of continuing education and training in cybersecurity. It says that remaining educated about the newest developments in cyber dangers and security measures is a constant effort. Regular training sessions and awareness programs for workers inside firms are vital in bolstering the human aspect, which is typically the first line of defense against cyber threats.

Finally, the chapter advises the cultivation of a resilient attitude in the face of cyber dangers. It accepts that even the most meticulous precautions, breaches may still occur. Therefore, companies must have contingency planning and crisis response protocols in place. This ability to promptly react to and recover from cyber disasters lowers possible harm and promotes a speedier return to regular operations.

In conclusion, "Preparing for Future Cyber Threats" is a critical chapter that highlights the essential need for diligent cybersecurity measures in our increasingly digital society. By identifying the dynamic nature of cyber threats, arguing for a multi-layered defense, fostering information exchange, prioritizing education, and highlighting resilience, this chapter prepares readers with a complete framework to defend against future cyber hazards. It serves as a timely reminder that in the digital era, proactive preparedness is crucial to ensuring a better and more secure future.

Chapter 8: Digital Inclusion and Accessibility

8.1. The Digital Divide: Global and Local Perspectives

"The Digital Divide: Global and Local Perspectives" is a crucial topic that tackles the discrepancies in access to and competency in information and communication technologies (ICTs) across various areas and people. This divide is defined by a substantial disparity between those who have quick access to digital tools and information, and those who do not. It involves not just the availability of physical infrastructure like computers and internet access, but also the capacity to successfully employ these technologies for educational, economic, and social goals.

On a worldwide basis, the digital gap is obvious in the different degrees of technology growth between industrialized and underdeveloped nations. Developed countries often claim high levels of internet penetration, modern communications networks, and extensive digital literacy. In contrast, underdeveloped nations generally suffer with inadequate access to dependable internet services, insufficient technical infrastructure, and lower rates of digital literacy. This mismatch in technology resources may considerably impair the socio-economic development of less affluent places, as access to knowledge and participation in the digital economy become more crucial for advancement.

At a local level, the digital gap may be seen inside particular nations, generally along socio-economic lines. Affluent metropolitan regions tend to have solid digital infrastructure, with broad access to high-speed internet and contemporary computer equipment. In contrast, rural and economically underprivileged regions may confront problems such as unstable connection, old technology, and limited possibilities for digital education and training. This local component of the digital divide exacerbates existing socio-economic disparities, since people without appropriate access to technology are at a severe disadvantage in terms of education, employment, and civic involvement.

Moreover, the digital divide goes beyond simple access to technology. It also covers the capacity to employ these resources successfully for personal and professional growth. Digital literacy, or the capacity to navigate and use digital platforms and apps, plays a vital role in bridging this difference. Individuals with greater degrees of digital literacy are more suited to seek jobs, access educational resources, and participate in the global information economy. Conversely, persons deficient in digital abilities suffer increased marginalization in an increasingly digitalized environment.

Efforts to overcome the digital gap need joint effort from governments, NGOs, and private sector parties. Initiatives aiming at improving access to technology, especially in disadvantaged places, are of critical significance. Additionally, educational initiatives focusing on digital literacy and skills development should be addressed to guarantee that persons from all walks of life are able to succeed in the digital age. Moreover, policies that stimulate innovation, competition, and investment in ICT infrastructure may go a long way in decreasing the global and local differences.

In conclusion, the digital divide is a complicated dilemma with global and local components. It involves inequities in access to technology, digital literacy, and the capacity to successfully employ ICTs for personal and professional growth. Addressing this gap is crucial for encouraging equitable socio-economic growth and ensuring that no person or community is left behind in the digital age. Through concentrated efforts and intelligent policies, it is feasible to bridge this difference and establish a more equal and empowered global community.

8.2. Ensuring Internet Access for All

"Ensuring Internet Access for All" is a vital component of the current digital world. In an era when information and services are largely accessible online, equal access to the internet has become a basic right, important for participation in education, work, civic engagement, and social inclusion.

First and foremost, widespread internet access is important for bridging the digital divide. Disparities in access based on socio-economic considerations, location, and demography remain, placing significant sectors of the community at a disadvantage. Without dependable internet connectivity, persons in underprivileged regions suffer limits in educational chances, career prospects, and accessing important services such as healthcare and government resources.

Moreover, the relevance of internet connection goes well beyond personal comfort. It plays a key role in democratizing knowledge and information. Through the internet, users have access to a wide collection of educational materials, ranging from online classes to open-access academic publications. This democratization of information levels the playing field and encourages everyone to seek learning and skills development irrespective of their geographical location or socio-economic position.

Additionally, internet connection is crucial in supporting economic growth. It helps entrepreneurs and small enterprises to contact global markets, offering potential for development and innovation. E-commerce platforms, for instance, give a platform for local craftspeople and small-scale manufacturers to expose their goods to a worldwide audience, therefore promoting economic activity at both local and international levels.

Furthermore, the provision of ubiquitous internet access encourages civic involvement and participation. It fosters contact between people and their governments, providing for the broadcast of information, feedback systems, and possibilities for public conversation. Access to government services and information online also simplifies bureaucratic procedures, making them more efficient and accessible to the larger populace.

However, it is vital to note that enabling internet access for everybody is not without its problems. Infrastructure development, especially in distant or economically challenged places, offers a considerable obstacle. Moreover, problems of price, digital literacy, and concerns about online safety and privacy must be addressed to guarantee that internet access is not just accessible but also inclusive and relevant for all parts of society.

In conclusion, "Ensuring Internet Access for All" is a vital endeavor in the digital era, with far-reaching consequences for education, economic growth, civic participation, and social inclusion. By addressing the inequities in access and working towards a more inclusive digital environment, we can harness the full potential of the internet as a force for good change in society. This undertaking involves collaborative efforts from governments, business sector institutions, and civil society to bridge the digital gap and build a more equal and connected world.

8.3. Assistive Technologies for Persons with Disabilities

Assistive technology have arisen as a transforming force in the lives of those with disabilities, allowing them increased freedom, empowerment, and accessibility. These technologies comprise a broad variety of equipment, software, and applications meant to lessen the difficulties that individuals with disabilities confront in their everyday life. One key component of assistive technology is its adaptation to multiple demands, addressing distinct sensory, mobility, cognitive, and communication issues. This article dives into the tremendous influence of assistive technology, emphasizing their role in creating inclusion and boosting the quality of life for individuals with disabilities.

One of the most remarkable triumphs of assistive technology is the enhanced autonomy they provide to those with impairments. For instance, mobility aids like wheelchairs and exoskeletons allow persons with mobility limitations to explore their surroundings with more comfort and freedom. Moreover, developments in voice recognition and assistive communication technology allow persons with speech and language challenges to express themselves effectively, exceeding the restrictions imposed by their diseases. By minimizing rely on others for fundamental activities, assistive technologies encourage a feeling of agency and self-reliance among persons with impairments.

Furthermore, assistive technology play a significant role in leveling the playing field for education and career prospects. Accessible educational software, screen readers, and adaptive learning aids enable students with impairments to participate fully in mainstream educational environments. Similarly, workplace accommodations like as screen magnifiers, ergonomic keyboards, and voice-activated software promote meaningful interaction in the professional world. These technologies not only boost productivity but also foster a more varied and inclusive workforce, using the unique talents and capabilities of those with disabilities.

In addition to addressing physical and sensory issues, assistive technology have made considerable progress in aiding those with cognitive impairments. For instance, customized software and applications enable persons with learning difficulties in skills like reading, writing, and organizing. Moreover, memory aides and task management applications support persons with cognitive impairments in everyday tasks, increasing their potential for independent living. By adapting solutions to particular cognitive demands, assistive technologies help people to overcome cognitive hurdles and attain their full potential.

Beyond individual advantages, assistive technology contribute to the greater social objective of inclusion. They serve as a key bridge between the digital world and those with impairments, ensuring that they may participate fully in the information age. This inclusion extends to social interactions, helping persons with communication challenges to connect more effectively with their peers and communities. By breaking down communication barriers, assistive technologies build a more inclusive and sympathetic society, where persons with disabilities are appreciated for their unique views and contributions.

In conclusion, assistive technology have altered the lives of those with disabilities, bringing them greater freedom, access to education and work, and prospects for social inclusion. These technologies illustrate the power of innovation in meeting the different requirements of those with disabilities, helping them to overcome physical, sensory, cognitive, and communication obstacles. As society continues to embrace inclusion and accessibility, the role of assistive technology will only grow more crucial in building a world where all persons, regardless of their ability, may prosper and contribute meaningfully.

8.4. Digital Literacy and Education

Digital literacy has become a vital ability in today's digitally driven culture. It involves the capacity to browse, assess, and use information in many digital formats, ranging from websites and social media to software programs and online databases. This expertise is vital not just for personal empowerment but also for success in school and the job. In the area of education, digital literacy plays a key role in designing current learning settings.

First and foremost, digital literacy empowers students with the ability to access a variety of knowledge and resources accessible on the internet. Through search engines, online libraries, and instructional platforms, learners may dig into a large variety of subjects, getting access to information that transcends geographical and institutional barriers. This democratization of knowledge encourages students to become active, self-directed learners, creating a feeling of autonomy and curiosity.

Moreover, digital literacy develops critical thinking and information assessment abilities. In an age when disinformation and false news spread, the ability to differentiate genuine sources from questionable ones is crucial. Students must learn how to evaluate material, determine its credibility, and confirm information across numerous sources. This not only assures the correctness of their work but also hones their analytical talents, helping them to make educated judgments in an increasingly complicated environment.

Additionally, digital literacy fosters cooperation and communication in school contexts. Online platforms, forums, and social media give outlets for students to participate in debates, exchange ideas, and cooperate on projects. This generates a feeling of community and allows varied viewpoints to come together, improving the learning experience. Moreover, it prepares students for the collaborative character of many professional situations, where excellent communication and cooperation are highly valued qualities.

However, problems exist in establishing universal digital literacy in education. Socioeconomic gaps may limit access to technology and high-quality internet connections, producing a digital divide that disproportionately impacts vulnerable areas. Efforts must be taken to guarantee equal access to digital materials, leveling the playing field for all learners. Additionally, educators require proper training and assistance in incorporating digital technologies successfully into their teaching methods, so as to maximize their advantages.

In conclusion, digital literacy is a cornerstone of contemporary education, giving students with the skills they need to prosper in a digitally-driven world. It allows students to access and critically assess information, stimulates cooperation and communication, and prepares them for the demands of a technology-intensive industry. While hurdles remain, coordinated efforts in providing access and training may bridge the digital divide and guarantee that all learners have the chance to build this vital skill set.

8.5. Closing the Gap: Policies and Initiatives

"Closing the Gap: Policies and Initiatives" emphasizes the crucial need for specific policies aimed at eliminating gaps in access to technology and digital resources. In an increasingly connected world, where access to information and online platforms is vital to participation in the global economy, such policies and efforts play a key role in guaranteeing inclusion. These activities span a variety of techniques, from government-led programs to private sector collaborations, all with the common objective of bridging the digital gap.

Government-led initiatives constitute the cornerstone of attempts to bridge the gap. These programs frequently include considerable expenditures in infrastructure, attempting to provide high-speed internet access to underprivileged populations. By dedicating resources towards creating powerful digital networks in historically neglected regions, governments attempt to level the playing field and equip residents with equitable chances for online involvement. Additionally, these policies may include subsidies or tax incentives to make technology more cheap, ensuring that socio-economic status does not limit access to important digital tools.

Public-private partnerships have emerged as a strong force in this quest. Collaborations between governments and IT businesses, for instance, have encouraged the construction of digital learning centers and community hubs. These venues act as entry points for persons who may not have the tools or skills to engage with technology on their own. Moreover, such agreements typically extend beyond physical places to include programs like subsidized equipment or free software licenses, further decreasing barriers to entry for individuals on the outskirts of digital access.

Educational efforts play a key role in equipping people to make the most of digital resources. Programs focusing on digital literacy give training and skills development to guarantee that people can traverse the online realm efficiently. From fundamental computer skills to more sophisticated digital capabilities, these projects empower individuals with the tools they need to effectively engage in the digital era. Furthermore, specialized instructional activities may also concentrate on certain populations, such as elders or those in distant places, who may face distinct hurdles in embracing digital technology.

Inclusive design concepts are a crucial component of narrowing the gap. By building technologies and platforms that are accessible to a varied spectrum of users, designers can guarantee that no one is mistakenly left behind. This comprises factors such as user-friendly interfaces, interoperability with assistive devices, and sensitivity to language and cultural diversity. Through such approaches, designers attempt to create an inclusive digital world where everyone can interact fully, regardless of background or ability.

In conclusion, "Closing the Gap: Policies and Initiatives" shows a holistic approach to solving the digital divide. Through government-led regulations, public-private collaborations, educational programs, and inclusive design principles, stakeholders try to create an environment where all persons have the chance to harness the promise of digital technology. By actively pursuing these tactics, society may move towards a more egalitarian and inclusive digital future, where access to technology is no longer a privilege but a basic right.

Chapter 9: Environmental Sustainability and Technology

9.1. Technology's Impact on the Environment

Technology's influence on the environment is a complicated and multidimensional issue, containing both good and bad impacts. On one side, technical breakthroughs have led to tremendous gains in resource efficiency and environmental conservation. For instance, the development of renewable energy technology like solar panels and wind turbines has decreased dependency on fossil fuels, lowering greenhouse gas emissions and addressing climate change. Moreover, advancements in waste management and recycling technology have lowered the ecological imprint of human activities.

However, it is vital to recognize the darker side of technology's effect on the environment. The high rate of urbanization and technological growth has resulted in widespread pollution and habitat devastation. The exploitation and usage of natural resources, typically driven by technology requirements, have led to deforestation, soil degradation, and disturbance of fragile ecosystems. The electronic waste created from old or abandoned technology constitutes a huge environmental problem, comprising toxic elements that may pollute land and water.

Furthermore, the manufacturing processes involved with technology production are usually energy-intensive and may contribute to air and water pollution. The mining of rare earth metals, crucial components in many high-tech products, has been connected to ecological devastation and significant health impacts for neighboring people. This highlights the complicated web of environmental repercussions woven into the fabric of contemporary technology.

In recent years, there has been a rising focus on sustainable technology development, aiming at reducing these unfavorable impacts. This entails developing items with a lifetime perspective, considering their environmental effect from manufacture to disposal. Additionally, eco-friendly materials and manufacturing processes are being studied to lessen the carbon footprint of technology. Regulations and standards have been created to ensure appropriate disposal and recycling of electronic trash, trying to reduce its adverse impacts on the environment.

Ultimately, the interaction between technology and the environment is a dynamic interplay, needing a balanced approach. While technology has the potential to transform environmental conservation efforts, it is vital that it be applied wisely and ethically. This demands a joint effort from governments, corporations, and people to prioritize sustainable practices and create new solutions that enable a peaceful coexistence between technology and the natural world. Only through this coordinated effort can we expect to pave the route towards a more environmentally-conscious technology future.

9.2. Green Technology and Renewable Energy

Green technology and renewable energy are essential components in the worldwide search of sustainable and ecologically responsible activities. This technological paradigm shift is motivated by an urgent need to alleviate the detrimental consequences of climate change and lessen dependence on limited fossil fuel supplies. At its heart, green technology comprises a wide range of technologies aimed to limit negative environmental consequences, concentrating on everything from energy generation and consumption to waste management and transportation.

Renewable energy, a major component of green technology, acts as a light of hope in the battle against climate change. Unlike traditional fossil fuels, renewable sources such as solar, wind, hydro, and geothermal power harness energy from naturally occurring and replenishable sources. The adoption of these technologies has undergone exponential development in recent years, driven by a mix of technology improvements, government incentives, and a rising social awareness of the need for sustainable energy options. Solar panels, for instance, have become symbolic of this transition, converting sunlight into a feasible and plentiful source of power.

One of the most enticing characteristics of green technology and renewable energy lies in their ability to drastically cut greenhouse gas emissions. By bypassing the burning of fossil fuels, these technologies restrict the emission of carbon dioxide and other pollutants into the atmosphere. This not only helps to prevent global warming but also mitigates the health hazards linked with air pollution. Moreover, the decentralization of energy generation, as evident in solar panels on household roofs or wind turbines in local towns, generates a feeling of energy independence and resilience, lowering susceptibility to centralized power outages.

In addition to environmental advantages, green technologies and renewable energy represent great economic possibilities. The move towards a green economy has led to the development of countless employment in production, installation, and maintenance of renewable energy systems. Furthermore, expenditures in research and development within this industry promote technical innovation, producing a positive feedback cycle of improvement. As countries internationally strive to diversify and modernize, the green technology sector is positioned to become a cornerstone of economic development and stability.

Nevertheless, problems exist in the mainstream use of green technologies and renewable energy. One important difficulty is the intermittency and unpredictability of various renewable sources. For instance, solar power production is depending on sunshine availability, whereas wind energy is dependent on constant wind patterns. To solve this, developments in energy storage technology and the development of smart grid systems are vital in maintaining a consistent and steady energy supply from renewable sources.

In conclusion, green technology and renewable energy constitute a vital crossroads in humanity's quest towards environmental sustainability. By harnessing the power of naturally replenishable resources and limiting environmental effect, these technologies provide a route towards a cleaner, more resilient energy future. Beyond environmental advantages, the integration of green technology has the potential to change economies and employment markets, moving humanity towards a more sustainable and wealthy future. As these technologies continue to advance and become more affordable, their broad adoption is expected to play a crucial role in minimizing the consequences of climate change and assuring a better future for generations to come.

9.3. Smart Cities and Sustainable Urban Planning

Smart cities offer a paradigm change in urban planning, merging technology, data, and connectivity to handle the complex difficulties of contemporary urbanization. At its heart, the idea envisions communities that leverage cutting-edge technology to boost efficiency, sustainability, and the general quality of life for citizens. Central to this concept is sustainable urban design, which plays a crucial role in guaranteeing the long-term sustainability and resilience of these cities.

Sustainable urban planning in the context of smart cities incorporates a comprehensive approach to development. It addresses variables such as energy efficiency, resource management, and the minimization of environmental effect. Planners are challenged with constructing areas that not only accommodate the rising population but also lessen the environmental burden associated with urbanization. This entails the integration of renewable energy sources, efficient waste management systems, and the promotion of green areas to maintain a healthy urban environment.

Furthermore, smart cities significantly depend on data-driven decision-making processes. Through the broad use of sensors, IoT devices, and real-time monitoring systems, cities may obtain a plethora of information on many elements of urban life. This data gives essential insights into the demands and habits of individuals, helping urban planners to make educated choices. For instance, it enables for the optimization of transportation networks, resulting to decreased congestion and fewer emissions.

Inclusivity is another cornerstone of sustainable urban development in smart cities. Planners must guarantee that the advantages of technology improvements are available to all sectors of the society. This covers factors for accessibility, cost, and digital literacy. By actively interacting with communities and integrating their views, urban planners may create places that respond to various needs and promote social fairness.

Resilience in the face of unanticipated difficulties, such as natural catastrophes or pandemics, is a vital part of sustainable urban design within smart cities. By using technology, communities may create early warning systems, efficient evacuation strategies, and strong healthcare infrastructure. This pre-emptive strategy not only preserves lives but also avoids economic and social disruptions during emergencies.

However, obstacles exist in the application of sustainable urban planning in smart cities. Balancing technical innovation with cost and scalability remains an issue. Additionally, the high speed of technology innovation needs adaptable design solutions that can keep pace with altering urban environments.

In conclusion, sustainable urban planning stands at the core of the smart city movement, pushing the creation of urban areas that are efficient, inclusive, and resilient. By merging modern technology, data-driven decision-making, and a dedication to environmental stewardship, cities may evolve into centers of innovation and sustainability. The success of this effort rests on the cooperation of urban planners, communities, and legislators, all working towards a shared vision of a brighter and more sustainable urban future.

9.4. The Circular Economy and Recycling in the Digital Age

"The Circular Economy and Recycling in the Digital Age" represents a key junction of environmental sustainability and technology innovation. This notion has gained relevance in recent years as civilizations cope with the rising issues of waste management and resource depletion. At its heart, the circular economy tries to reinvent the old linear paradigm of "take-make-dispose" and substitutes it with a more holistic approach where resources are continually cycled back into the manufacturing process. In the context of the digital era, technology plays a vital role in facilitating and boosting the efficiency of circular economy processes.

Digital technology are crucial in transforming trash management and recycling operations. Advanced sensors, Internet of Things (IoT) devices, and data analytics allow the monitoring and tracking of resources throughout their lifespan. For instance, smart bins connected with sensors may offer real-time data on fill levels, improving collection routes and minimizing needless trips. This not only conserves gasoline but also mitigates pollutants, harmonizing with the larger aims of environmental sustainability.

Moreover, the digital era has produced unique technologies for material identification and sorting. Machine learning algorithms, in conjunction with high-resolution photography, allow recycling plants to effectively sort various materials. This reduces contamination and enhances the quality of recycled goods, making them more competitive with their virgin counterparts. Additionally, blockchain technology is developing as a method to construct transparent supply chains for recycled materials, confirming their origin and validity.

E-commerce platforms and digital marketplaces play a crucial part in the circular economy ecosystem. These platforms promote the resale and redistribution of items, prolonging their lives and lowering the necessity for new manufacture. Online markets for pre-owned items, backed by user ratings and reviews, develop trust among customers, therefore creating a culture of reuse. Furthermore, digital platforms give a chance for firms to embrace new business models such as product-as-a-service and sharing economies, therefore enhancing the value of items and eliminating waste.

Education and awareness are key components of establishing a circular economy in the digital era. Technology-enabled communication channels, such as social media and online forums, serve as strong instruments for spreading knowledge and effecting behavioral change. Campaigns and initiatives pushing for recycling and sustainable consumerism may reach a larger audience and involve communities in a joint effort towards a more sustainable future. Additionally, digital platforms may offer essential materials, such as tutorials and guidelines, to assist people and organizations adopt more sustainable practices.

In conclusion, the combination of the circular economy and digital technology provides a viable road towards sustainable resource management in the digital era. Technology supports the optimization of waste management operations, promotes material identification and sorting, and allows the construction of transparent supply chains. E-commerce platforms play a vital role in encouraging the reuse and redistribution of items, while digital media act as effective instruments for education and awareness. As we traverse the difficulties of a fast developing world, exploiting the potential of the circular economy in connection with digital innovation is not just a pragmatic answer but also a visionary step towards a more sustainable and resilient future.

9.5. Climate Change Mitigation through Technology

Climate change is one of the most important global concerns of our day, with far-reaching environmental, social, and economic ramifications. As the globe grapples with the necessity of lowering greenhouse gas emissions, technology emerges as a vital instrument in the battle against climate change. The notion of climate change mitigation via technology comprises a spectrum of creative solutions targeted at limiting the consequences of human activities on the planet's climatic systems.

One of the important parts of climate change mitigation via technology is in the development and deployment of renewable energy sources. Transitioning from fossil fuels to renewable energy is crucial to lowering carbon emissions. Solar, wind, hydro, and geothermal energy technologies have experienced considerable breakthroughs in recent years, making them more viable alternatives to traditional, carbon-intensive energy sources. These technologies not only supply sustainable energy but also contribute to the decarbonization of numerous sectors, including transportation, industrial, and residential.

Additionally, developments in energy efficiency technology play a key part in climate change mitigation efforts. From increased building insulation to more energy-efficient appliances and industrial processes, technological improvements have considerably strengthened our potential to do more with less energy. The introduction of smart grids and the integration of IoT (Internet of Things) technology further improve energy consumption patterns, allowing for real-time changes and decreased waste. By adopting these technologies, towns and companies may minimize their carbon footprint while also saving on energy expenditures.

Furthermore, advancement in carbon capture and storage (CCS) technology offers enormous potential for climate change mitigation. CCS includes absorbing carbon dioxide emissions generated from the use of fossil fuels in energy production and industrial activities. These caught emissions are then transported and stored underground, keeping them from exiting the atmosphere and contributing to global warming. While still in the early phases of research and implementation, CCS has the potential to play a key role in lowering emissions from industries where total decarbonization is a tough undertaking.

The introduction of circular economy technology also provides a significant method to climate change mitigation. This paradigm shift stresses the effective use of resources, materials, and products, reducing waste and increasing recycling and reuse. From improved recycling technology to the production of bio-based products, the circular economy links economic progress with environmental sustainability. By decreasing the need for fresh resource extraction and eliminating waste creation, this technique helps greatly to cutting greenhouse gas emissions.

In conclusion, climate change mitigation via technology is a vital pillar in our collaborative efforts to counteract global warming and its accompanying repercussions. The development and deployment of renewable energy sources, breakthroughs in energy efficiency, carbon capture and storage technologies, and the adoption of circular economy ideas all play crucial roles in this quest. By leveraging the power of technology, we have the capacity to shift towards a more sustainable and resilient future, where economic success coexists happily with environmental well-being.

Chapter 10: The Future of Technology and Human Society

10.1. Anticipating Future Technological Trends

"Anticipating Future Technological Trends" is a vital component of navigating the continually moving environment of technology and innovation. This component entails understanding the trajectories and possible influence of developing technologies on diverse sectors, society, and ordinary life. It comprises the capacity to forecast developments and modify methods to use them successfully. Anticipation frequently comprises a mix of data analysis, trend monitoring, and expert insights.

One crucial part of projecting future technology developments is understanding the convergence of present technologies. This happens when diverse technical fields, such as artificial intelligence, blockchain, and IoT, meet to provide unique solutions. For example, the integration of AI algorithms into healthcare equipment has led to the creation of individualized treatment programs and better patient outcomes. Identifying these convergences helps organizations and sectors to put themselves at the forefront of innovation.

Moreover, recognizing the socio-economic ramifications of developing technology is crucial. As technology rapidly enters all elements of everyday life, it brings about transformations in employment, education, and society institutions. For instance, the advent of automation and AI-driven systems has generated questions about worker reskilling and the nature of employment in the future. Organizations and governments need to proactively address these changes to guarantee a seamless transition and prevent any disruptions.

Another key part of forecasting technology trends is recognizing the ethical and regulatory implications connected with emerging technologies. The advent of advancements like gene editing, driverless cars, and quantum computing presents fundamental ethical issues. Balancing technical advancement with ethical bounds needs significant debate and the construction of solid regulatory systems. Anticipation, in this context, requires not just knowing the prospective advantages but also being aware of the ethical problems and social ramifications.

Furthermore, keeping an eye on developing industries and startups is vital for staying ahead of technology developments. Often, disruptive breakthroughs arise from startups and less established firms in the business. Understanding their methodologies and technology may give useful insights into the trajectory of the technological world. Collaborative actions, like as alliances or acquisitions, with these businesses may be important in leveraging the potential of these developing trends.

In conclusion, "Anticipating Future technical Trends" is a multidimensional activity that demands a complete awareness of technical landscapes. This involves spotting convergences, resolving socio-economic repercussions, managing ethical issues, and connecting with new markets. Proactive foresight not only helps companies and sectors to adapt to change but also positions them to create the future of technology and innovation. It is a vital part of flourishing in the dynamic digital era.

10.2. Ethical Considerations in Technological Development

"Ethical Considerations in technology Development" spans a vital feature of the contemporary world, stressing the enormous influence that technology breakthroughs have on society, people, and the environment. This topic dives into the complicated interaction between innovation and morality, inviting profound contemplation on how technology should be handled to guarantee a good and sustainable future.

One of the primary ethical issues focuses around privacy and data protection. As technology continues to improve, the gathering and broadcast of personal data have become omnipresent. Companies and organizations increasingly store huge troves of information about people, prompting questions about permission, transparency, and possible exploitation. Striking a balance between the advantages of data-driven innovation and maintaining individual privacy is a continuous difficulty in current technology progress.

Moreover, inclusion and accessibility are important ethical features of technology growth. It's vital that technology be built with the purpose of assisting all members of society, regardless of age, aptitude, or socio-economic background. Neglecting this concept may lead to digital divides, where some people are excluded from the advantages and possibilities that technology gives. Ethical developers, then, must emphasize designing solutions that are accessible to varied audiences, guaranteeing equal access to the rewards of technical innovation.

Additionally, a major ethical dilemma refers to the larger social implications of technology. Innovations, although potentially helpful, may also undermine existing institutions and lives. Ethical technologists need to rigorously examine and minimize possible harmful repercussions. This requires addressing the socio-economic repercussions, possible job displacement, and ensuring that technology uplifts rather than marginalizes populations.

Environmental sustainability is yet another essential ethical problem. As the need for technology rises, so too does its ecological imprint. From resource extraction to energy use, technological advancement must be undertaken with an understanding of its environmental implications. Ethical concerns necessitate a commitment to sustainable activities, including the ethical use of resources and the creation of eco-friendly technology.

Lastly, the notion of openness and accountability is crucial to ethical technology development. Users and society at large must have a comprehensive grasp of how technologies function, and the possible repercussions of their usage. Companies and developers hold an obligation to be upfront about their objectives, procedures, and any possible hazards linked with their goods. Moreover, procedures for accountability must be in place to address any unforeseen or adverse outcomes that may develop.

In conclusion, ethical concerns in technological progress are a critical component in building a future where technology promotes the well-being of mankind and the earth. Balancing innovation with morality needs a holistic strategy that incorporates privacy, inclusion, social effect, environmental sustainability, and openness. Embracing these ethical imperatives ensures that technology advancement corresponds with the goals and ambitions of a fair and flourishing global society.

10.3. Balancing Innovation with Human Well-being

"Balancing Innovation with Human Well-being" is a critical part of managing the quickly expanding environment of technology and its consequences on society. This concept emphasizes that although innovation drives development and provides a plethora of advantages, it must be balanced with a thorough evaluation of its consequences on persons and society. Striking this balance guarantees that technology innovations do not unwittingly injure or exploit humans, and that they contribute positively to human happiness.

At its root, reconciling innovation with human well-being needs a careful approach towards technology advancement. It entails an ethical duty on the part of inventors, companies, and legislators to examine the larger social ramifications of their products. This includes assessing not just the possible advantages and economic gains but also the potential concerns, particularly those relating to privacy, security, and the equal distribution of benefits.

Furthermore, this concept fosters the incorporation of inclusiveness and accessibility in the design and implementation of new technologies. Innovations should not mistakenly alienate or exclude some groups of the community. This requires recognizing the needs of various populations, especially individuals with impairments or limited access to resources. By doing so, technology may become a vehicle for inclusiveness, empowerment, and equality, rather than increasing existing gaps.

Balancing innovation with human well-being also needs the construction of solid regulatory systems. These frameworks should attempt to defend the rights, privacy, and dignity of persons while yet allowing for the development and deployment of transformational technology. Striking this regulatory balance is vital to avoid possible abuses, such as the unethical use of personal data or the unrestrained spread of potentially hazardous technology.

Moreover, this approach advocates a proactive posture towards predicting and reducing unforeseen effects of technology progress. This may include continual monitoring, iterative development, and a readiness to adjust when new knowledge and social implications emerge. It recognizes that our understanding of technology's consequences is dynamic and that responsible innovation needs a constant commitment to learning and adaptation.

In conclusion, "Balancing Innovation with Human Well-being" is a key premise in ensuring that technology advancement serves the good of mankind rather than its harm. It stresses ethical issues, inclusion, regulatory frameworks, and a proactive approach towards resolving possible dangers. By adhering to this idea, we may build a future where innovation and human well-being live together, leading to a brighter and more egalitarian digital era.

10.4. Preparing for Technological Challenges

"Preparing for Technological Challenges" comprises the tactics and techniques people, companies, and society apply to foresee and alleviate possible challenges coming from the fast growth of technology. This chapter digs into the proactive actions that may be done to establish a resilient environment in the face of ever-evolving digital landscapes.

Firstly, a significant component of preparing for technological challenges entails developing a culture of adaptation and continual learning. In an age when new technologies develop at an unprecedented rate, people and organizations must foster a culture that prioritizes lifelong learning. This involves maintaining informed with the newest trends, attending seminars or training sessions, and developing a culture of inquiry and experimentation. By doing so, stakeholders position themselves to not only keep up with technology changes but also harness them effectively for innovation and problem-solving.

Secondly, effective cybersecurity measures are crucial in preparing for technological problems. With the development of cyber dangers, from sophisticated hacking tactics to ransomware assaults, securing sensitive information has become a major issue. Implementing encryption techniques, multi-factor authentication, and regular security audits are just a few measures to improve digital defenses. Furthermore, teaching users about cyber hygiene habits is as crucial, since the human component remains a key threat. By establishing a security-conscious culture, companies may considerably lower their vulnerability to cyber attacks.

Another crucial component of preparation for technological challenges is ethical considerations. As technologies like artificial intelligence and machine learning become more incorporated into our everyday lives, it becomes important to build ethical frameworks that regulate their usage. This involves tackling concerns like algorithmic bias, data privacy, and transparency. Additionally, building forums for public conversation and policymaking ensures that technical improvements match with society ideals and prevent possible hazards.

Furthermore, effective disaster recovery and contingency planning are key components of preparing for technology difficulties. Given the interdependence of digital systems, unplanned occurrences such as system breakdowns, natural catastrophes, or cyber-attacks may have far-reaching implications. Establishing redundancy in important systems, frequent backups, and well defined response methods are vital in avoiding downtime and data loss. These methods not only boost resilience but also inspire trust in stakeholders that their digital infrastructure can endure unanticipated crises.

In conclusion, "Preparing for Technological Challenges" comprises a holistic strategy that combines a proactive mentality, cybersecurity safeguards, ethical concerns, and disaster preparation. Embracing a culture of constant learning and flexibility, together with rigorous security standards, is the underpinning of preparedness. Additionally, incorporating ethical standards and building contingency plans further hardens the basis. By employing these techniques, people, businesses, and society may manage the developing technology environment with confidence and resilience, assuring a better digital future.

10.5. A Vision for a Brighter Digital Future

"A Vision for a Brighter Digital Future" reflects the collective goal for a technologically evolved future that harmoniously merges innovation, inclusion, and ethical issues. This last chapter serves as both a conclusion and a call to action, mapping a route towards a digital world that enhances human existence. It underlines the vital significance of balancing between fast technological growth and the preservation of our common humanity.

In visualizing a better digital future, one fundamental topic emerges: the requirement of ethical concerns in technical growth. The chapter highlights that technical advancement must be directed by a strong ethical compass, ensuring that advances are not simply about capabilities, but also about the good of society. It challenges us to ponder on problems of privacy, autonomy, and social justice, challenging us to favor technologies that elevate and empower people rather than exploit or marginalize them.

Furthermore, the chapter pushes for diversity as a cornerstone of this envisioned future. It underscores the importance to bridge the digital divide, ensuring that all persons, regardless of background or situation, have access to the possibilities and advantages given by technology. This inclusion goes beyond simple access to incorporate digital knowledge and education, realizing that genuine empowerment emerges from the capacity to harness technology for personal and community advancement.

A significant element of this strategy centers upon environmental sustainability. The chapter underlines the necessity for technology to be utilized in the service of ecological protection. It envisions a future where technology is not just a source of economic progress but also a tool for generating sustainable solutions to critical environmental concerns. Concepts like green technology and the circular economy are put up as crucial components in fulfilling this goal.

The chapter doesn't shy away from the hurdles involved in this endeavour. It recognizes the possible perils and downsides of uncontrolled technological growth, cautioning against complacency or short-sightedness. It supports continuing conversation, research, and policy-making to lead technology towards a future that maximizes benefits while reducing damage.

Ultimately, "A Vision for a Brighter Digital Future" is a call to common accountability. It compels us to grasp that the design of the digital future is not preordained, but rather a consequence of our choices, actions, and ideals. It highlights that, with intentional effort and ethical foresight, we possess the capacity to construct a future where technology is a force for good, enhancing the lives of people and communities globally. This vision is a reminder that, although technology is a powerful instrument, it is our common humanity that ultimately defines its influence on the world.

Refrences

1.	Anderson, C. (2008). The Long Tail: Why the Future of Business is Selling Less of More. Hyperion.
2.	Barabási, A. L. (2003). Linked: How Everything Is Connected to Everything Else and What It Means for Business, Science, and Everyday Life. Plume.
3.	Castells, M. (1996). The Rise of the Network Society. Blackwell.
4.	Davenport, T. H., & Harris, J. (2007). Competing on Analytics: The New Science of Winning. Harvard Business Review Press.
5.	Floridi, L. (2010). Information: A Very Short Introduction. Oxford University Press.
6.	Friedman, T. L. (2006). The World Is Flat: A Brief History of the Twenty-First Century. Farrar, Straus and Giroux.
7.	Gates, B. (1995). The Road Ahead. Viking.
8.	Giddens, A. (1990). The Consequences of Modernity. Stanford University Press.
9.	Johnson, S. (2010). Where Good Ideas Come From: The Natural History of Innovation. Riverhead Books.
10.	Kelly, K. (2010). What Technology Wants. Viking.
11.	Lessig, L. (2008). Remix: Making Art and Commerce Thrive in the Hybrid Economy. Penguin Press.

12. Minsky, M. (1986). The Society of Mind. Simon & Schuster.

13. Negroponte, N. (1995). Being Digital. Vintage.

14. Pink, D. H. (2009). Drive: The Surprising Truth About What Motivates Us. Riverhead Books.

15. Rifkin, J. (2014). The Zero Marginal Cost Society: The Internet of Things, the Collaborative Commons, and the Eclipse of Capitalism. St. Martin's Press.

16. Shirky, C. (2008). Here Comes Everybody: The Power of Organizing Without Organizations. Penguin Press.

17. Tapscott, D., & Williams, A. D. (2006). Wikinomics: How Mass Collaboration Changes Everything. Penguin.

18. Toffler, A. (1980). The Third Wave. Bantam Books.

19. Turkle, S. (2011). Alone Together: Why We Expect More from Technology and Less from Each Other. Basic Books.

20. Weinberger, D. (2011). Too Big to Know: Rethinking Knowledge Now That the Facts Aren't the Facts, Experts Are Everywhere, and the Smartest Person in the Room Is the Room. Basic Books.

21. Anderson, P. (2008). Free: The Future of a Radical Price. Hyperion.

22. Benkler, Y. (2006). The Wealth of Networks: How Social Production Transforms Markets and Freedom. Yale University Press.

23. Berman, S. J. (2012). Digital Innovation: The Rise and Fall of Information Empires. Morgan Kaufmann.

24. Christensen, C. M. (1997). The Innovator's Dilemma: When New Technologies Cause Great Firms to Fail. Harvard Business Review Press.

25. Crawford, S. L., & Mathews, M. C. (2008). Introduction to Work in the 21st Century: An Anthology of Current Workplace Practices. Ballantine Books.

26. Evans, P., & Wurster, T. S. (2000). Blown to Bits: How the New Economics of Information Transforms Strategy. Harvard Business Review Press.

27. Gladwell, M. (2000). The Tipping Point: How Little Things Can Make a Big Difference. Little, Brown, and Company.

28. Godin, S. (1999). Permission Marketing: Turning Strangers into Friends, and Friends into Customers. Simon & Schuster.

29. Kelly, K. (1994). Out of Control: The New Biology of Machines, Social Systems, and the Economic World. Addison-Wesley.

30. Lanier, J. (2010). You Are Not a Gadget: A Manifesto. Vintage.

31. Lévy, P. (1997). Collective Intelligence: Mankind's Emerging World in Cyberspace. Perseus Books.

32. Li, C., & Bernoff, J. (2008). Groundswell: Winning in a World Transformed by Social Technologies. Harvard Business Review Press.

33. McChesney, R. W. (1999). Rich Media, Poor Democracy: Communication Politics in Dubious Times. University of Illinois Press.

34. Pinker, S. (2018). Enlightenment Now: The Case for Reason, Science, Humanism, and Progress. Viking.

35. Rheingold, H. (2002). Smart Mobs: The Next Social Revolution. Basic Books.

36. Rushkoff, D. (2010). Program or Be Programmed: Ten Commands for a Digital Age. OR Books.

37. Shirky, C. (2010). Cognitive Surplus: Creativity and Generosity in a Connected Age. Penguin Press.

38. Surowiecki, J. (2004). The Wisdom of Crowds. Anchor Books.

39. Thrun, S., & Norvig, P. (2009). Artificial Intelligence: A Modern Approach. Pearson.

40. Weinberger, D. (2007). Everything Is Miscellaneous: The Power of the New Digital Disorder. Times Books.